The Handbook of Environmental Chemistry

Founded by Otto Hutzinger

Editors-in-Chief: Damià Barceló · Andrey G. Kostianoy

Volume 10

The Handbook of Environmental Chemistry
Recently Published and Forthcoming Volumes

Chlorinated Paraffins

Volume Editor: Jacob de Boer

With contributions by

T. El-Sayed Ali · H. Fiedler · J. Legler · D. Muir ·
V.A. Nikiforov · G.T. Tomy · K. Tsunemi

 Springer

Editor
Prof. Dr. Jacob de Boer
Department of Chemistry and Biology
Institute for Environmental Studies (IVM)
VU University
De Boelelaan 1087
1081 HV Amsterdam
The Netherlands
jacob.de.boer@ivm.vu.nl

The Handbook of Environmental Chemistry
ISSN 1867-979X   e-ISSN 1616-864X
ISBN 978-3-642-10760-3   e-ISBN 978-3-642-10761-0
DOI 10.1007/978-3-642-10761-0
Springer Heidelberg Dordrecht London New York

Library of Congress Control Number: 2010927758

Cover design: SPi Publisher Services

Printed on acid-free paper

Springer is part of Springer Science+Business Media (www.springer.com)

The Handbook of Environmental Chemistry
Also Available Electronically

The Handbook of Environmental Chemistry is included in Springer's eBook package *Earth and Environmental Science*. If a library does not opt for the whole package, the book series may be bought on a subscription basis.

For all customers who have a standing order to the print version of *The Handbook of Environmental Chemistry,* we offer the electronic version via SpringerLink free of charge. If you do not have access, you can still view the table of contents of each volume and the abstract of each article on SpringerLink (www.springerlink.com/content/110354/).

You will find information about the

– Editorial Board
– Aims and Scope
– Instructions for Authors
– Sample Contribution

at springer.com (www.springer.com/series/698).

All figures submitted in color are published in full color in the electronic version on SpringerLink.

Aims and Scope

Since 1980, *The Handbook of Environmental Chemistry* has provided sound and solid knowledge about environmental topics from a chemical perspective. Presenting a wide spectrum of viewpoints and approaches, the series now covers topics such as local and global changes of natural environment and climate; anthropogenic impact on the environment; water, air and soil pollution; remediation and waste characterization; environmental contaminants; biogeochemistry; geo-ecology; chemical reactions and processes; chemical and biological transformations as well as physical transport of chemicals in the environment; or environmental modeling. A particular focus of the series lies on methodological advances in environmental analytical chemistry.

Series Preface

With remarkable vision, Prof. Otto Hutzinger initiated *The Handbook of Environmental Chemistry* in 1980 and became the founding Editor-in-Chief. At that time, environmental chemistry was an emerging field, aiming at a complete description of the Earth's environment, encompassing the physical, chemical, biological, and geological transformations of chemical substances occurring on a local as well as a global scale. Environmental chemistry was intended to provide an account of the impact of man's activities on the natural environment by describing observed changes.

While a considerable amount of knowledge has been accumulated over the last three decades, as reflected in the more than 70 volumes of *The Handbook of Environmental Chemistry,* there are still many scientific and policy challenges ahead due to the complexity and interdisciplinary nature of the field. The series will therefore continue to provide compilations of current knowledge. Contributions are written by leading experts with practical experience in their fields. *The Handbook of Environmental Chemistry* grows with the increases in our scientific understanding, and provides a valuable source not only for scientists but also for environmental managers and decision-makers. Today, the series covers a broad range of environmental topics from a chemical perspective, including methodological advances in environmental analytical chemistry.

In recent years, there has been a growing tendency to include subject matter of societal relevance in the broad view of environmental chemistry. Topics include life cycle analysis, environmental management, sustainable development, and socio-economic, legal and even political problems, among others. While these topics are of great importance for the development and acceptance of *The Handbook of Environmental Chemistry,* the publisher and Editors-in-Chief have decided to keep the handbook essentially a source of information on "hard sciences" with a particular emphasis on chemistry, but also covering biology, geology, hydrology and engineering as applied to environmental sciences.

The volumes of the series are written at an advanced level, addressing the needs of both researchers and graduate students, as well as of people outside the field of "pure" chemistry, including those in industry, business, government, research establishments, and public interest groups. It would be very satisfying to see these volumes used as a basis for graduate courses in environmental chemistry. With its high standards of scientific quality and clarity, *The Handbook of Environmental*

Chemistry provides a solid basis from which scientists can share their knowledge on the different aspects of environmental problems, presenting a wide spectrum of viewpoints and approaches.

The Handbook of Environmental Chemistry is available both in print and online via www.springerlink.com/content/110354/. Articles are published online as soon as they have been approved for publication. Authors, Volume Editors and Editors-in-Chief are rewarded by the broad acceptance of *The Handbook of Environmental Chemistry* by the scientific community, from whom suggestions for new topics to the Editors-in-Chief are always very welcome.

Damià Barceló
Andrey G. Kostianoy
Editors-in-Chief

Volume Preface

In the twentieth century, the combination of an exploding world population and an ongoing industrial revolution caused an entirely new hazard for mankind: exposure to a large variety of chemicals that are present in our food, drinking and swimming water, and in the air. In particular, halogenated organic compounds have a persistent character. They are not easily degraded, accumulate in organisms such as birds, fishes and marine mammals and enter the human body where, dependent of their concentration, they cause an array of effects such as immunotoxic, carcinogenic and endocrine disrupting effects. Nowadays a number of these chemicals have been regulated in many countries, often through production bans such as for polychlorinated biphenyls (PCBs) and toxaphene. Most of them are, however, still around and their environmental levels are being monitored in national and international programmes such as the Stockholm Convention on Persistent Organic Pollutants (POPs).

Remarkably, the class of chlorinated paraffins has not received much attention, while the worldwide production surpasses by far that of PCBs and substantial environmental levels are being reported. The enormous complexity of the mixture with tens of thousands of congeners may be one of the reasons for that. Environmental analysts have been deterred by this complexity, which demands qualitative and quantitative methods at a level beyond the performance characteristics of their instruments. However, now the selectivity and sensitivity of analytical instruments are improving rapidly and have reached performance levels that were previously unthought-of. That brings a reliable analysis of chlorinated paraffins within reach, though it is still not easy.

This book describes the state-of-the-art methods for synthesis and analysis of chlorinated paraffins. It provides an overview of their worldwide occurrence and impact and describes their toxicological properties. International regulations and production volumes are presented, as well as an example of a risk assessment study that was carried out in Japan. Therefore, this book will be useful not only for environmental scientists who need to study the occurrence and toxicology of chlorinated paraffins in environmental matrices, but also for authorities and producers who could use this book as a valuable and comprehensive source of information.

Chlorinated paraffins are normally divided into three sub-groups: short-chain (SCCP), medium-chain (MCCP) and long-chain (LCCP) chlorinated paraffins.

The emphasis in this book is on SCCPs, as most of the information available is on this sub-group. SCCPs have carbon chains of C_{10}-C_{13}. However, some chapters have included valuable information on MCCPs (C_{14}-C_{17}) and LCCPs (C_{20}-C_{30}) as well. We have used the term "chlorinated paraffins" throughout the book, whereas "chlorinated alkanes" is also used in the literature.

Amsterdam   Jacob de Boer
February 2010

Contents

Short-Chain Chlorinated Paraffins: Production, Use and International Regulations

Heidelore Fiedler

Abstract Chlorinated paraffins (CPs) are a group of synthetic organic chemicals consisting of n-alkanes with varying degrees of chlorination, usually between 40 and 70% by weight. There are no known natural sources of CPs. CPs are produced by chlorination of n-alkane feedstocks. CPs typically are viscous oils with low vapor pressures; they are practically insoluble in water but are soluble in chlorinated solvents or mineral oils. They are toxic to wildlife, long-lasting in the environment and build up in the tissues of organisms. Long-chain CPs are believed to be much less toxic to aquatic life than the related short- or medium-chain CPs.

CPs consist of extremely complex mixtures allowing many possible positions for the chlorine atoms. Depending on the degree of chlorination, they are grouped into low ($<$50%) and high ($>$50%) chlorine containing. Depending on the chain length, the products are often subdivided into short-chain (C_{10}–C_{13}), medium-chain (C_{14}–C_{17}) and long-chain (C_{18}–C_{30}) CPs.

CPs, including short-chain chlorinated paraffins (SCCPs), are used worldwide in a wide range of applications such as plasticisers in plastics, extreme pressure additives in metalworking fluids, flame retardants and additives in paints. Their wide industrial applications probably provide the major source of environmental contamination. CPs may be released into the environment from improperly disposed metalworking fluids containing CPs or from polymers containing CPs. Loss of CPs by leaching from paints and coatings may also contribute to environmental contamination. The potential for loss during production and transport is expected to be less than that during product use and disposal. Despite many efforts, a global picture as to the definition of CPs, present production, uses and occurrences is still not yet obtained.

Since about 20 years, SCCPs have become subject to regulation at national and international level due to their physical–chemical properties and adverse effects.

H. Fiedler
UNEP Chemicals, 11-13, Chemin des Anémones, CH-1219 Châtelaine (GE), Switzerland
e-mail: heidelore.fiedler@unep.org

J. de Boer (ed.), *Chlorinated Paraffins*, Hdb Env Chem (2010) 10: 1–40,
DOI 10.1007/698_2010_58, © Springer-Verlag Berlin Heidelberg 2010,
Published online: 14 April 2010

Action has been initiated for severely restricting or banning production and use of certain CPs. The latest activities include the listing of SCCPs under the Persistent Organic Pollutants (POPs) Protocol of the United Nations Economic Commission for Europe (UNECE) Longe-Range Transboundary Air Pollution Convention and ongoing discussions on including SCCPs to the Stockholm Convention on POPs.

Keywords Chlorinated paraffins, Definitions, Regulation, Risk assessment, Releases

Contents

Abbreviations

ACP	Arctic contamination potential
b.w.	Body weight
CAS	Chemical Abstract Service
COP	Conference of the parties
CP(s)	Chlorinated paraffin(s)
CSTEE	European Scientific Committee on Toxicity, Ecotoxicity and the Environment
CTV	Critical toxicity value
E(E)C	European (Economic) Commission
EEV	Estimated exposure value
EINECS	European Inventory of Existing Chemical Substances

EPER	European Pollutant Emission Register
E-PRTR	European Pollutant Release and Transfer Register
GMP	Global monitoring plan
HELCOM	Helsinki Commission
HLC	Henry's law constant
IARC	International Agency for Research on Cancer
IPCS	International Programme on Chemical Safety
IUPAC	International Union for Applied Chemistry
LCCPs	Long-chain chlorinated paraffins
LOAEL	Lowest observed adverse effect level
LOEC	Lowest observed effect concentration
LRTAP	Long-range transboundary air pollution
MAP	Mediterranean Action Plan
MCCPs	Medium-chain chlorinated paraffins
NOAEL	No adverse effect level
OECD	Organisation for Economic Co-operation and Development
OSPAR	Oslo-Paris Convention for the Protection of the Marine Environment of the North-East Atlantic
PARCOM	Paris Commission
PBT	Persistent, bioaccumulative and toxic
PEC	Predicted environmental concentration
PNEC	Predicted no-effect concentration
POPRC	Persistent Organic Pollutants Review Committee
POPs	Persistent organic pollutants
PVC	Polyvinylchloride
RAR	(European) Risk Assessment Report
RQ	Risk quotient
SCCPs	Short-chain chlorinated paraffins
STP	Sewage treatment plant
TDI	Tolerable daily intake
UNECE	United Nations Economic Commission for Europe
UNEP	United Nations Environment Programme
WFD	Water Framework Directive
WHO	World Health Organisation

1 Background

Chlorinated paraffins (CPs) or chlorinated *n*-alkanes are a group of synthetic compounds produced by the chlorination of straight-chained paraffin fractions. The feedstock used determines the carbon chain length distribution of the product. The carbon chain length of commercial CPs is usually between 10 and 30 carbon atoms, and the chlorine content is usually between 40% and 70% by weight. In general, there are three different carbon-chain length feedstocks that are used to manufacture CPs: short-chain (C_{10}–C_{13}), medium-chain (C_{14}–C_{17}) and long-chain (C_{18}–C_{30}).

Short-chain (SCCPs), medium-chain (MCCPs), and long-chain (LCCPs) CPs all have industrial applications and similar environmental concerns. CPs are viscous, colorless or yellowish dense oils with low vapor pressures, except for those of long carbon chain length with high chlorine content (70%), which are solid. CPs are practically insoluble in water, lower alcohols, glycerol and glycols, but are soluble in chlorinated solvents, aromatic hydrocarbons, ketones, esters, ethers, mineral oils and some cutting oils. They are moderately soluble in unchlorinated aliphatic hydrocarbons [1].

They tend to be oily liquids or waxy solids and they are toxic to wildlife, long-lasting and build up in the tissues of organisms. LCCPs are believed to be much less toxic to aquatic life than the related SCCPs and MCCPs [2].

Briefly, LCCPs and MCCPs occur as complex mixtures; they are waxy solids that decompose at over 200°C with the release of hydrogen chloride gas. They are virtually insoluble in water but dissolve fully in most non-polar organic solvents like paraffin oil. They are non-flammable and do not evaporate easily. The main use of LCCPs and MCCPs is as plasticisers in flexible polyvinylchloride (PVC), industrial metalworking fluids used in processes involving cutting, drilling, machining and stamping metal in engineering and manufacturing. Because of their fire retarding properties, they are sometimes added to rubbers for particular applications. They are also used in paints and other coatings, leathers, textiles and sealing compounds.

CPs are non-biodegradable and are bioaccumulative. They have been detected in marine and freshwater animals and in sediments in industrial areas and have also been found in remote locations. They are toxic to aquatic life.

SCCPs and MCCPs are a priority for risk assessment under the Council Regulation EEC 793/93 of 23 March 1993. CPs are candidates for selection, assessment and prioritisation under the OSPAR strategy and priority chemicals under the European Water Framework Directive (WFD).

SCCPs are mainly used in metalworking fluids, in sealants, as flame retardants in rubbers and textiles, in leather processing and in paints and coatings. SCCPs having carbon chain lengths of between 10 and 13 carbon atoms have a degree of chlorination of more than 48% by weight. Commercially available C_{10}–C_{13} CPs are usually mixtures of different carbon chain lengths and different degrees of chlorination, although all have a common structure, in that no secondary carbon atom carries more than one chlorine atom.

Since SCCPs are highly toxic to aquatic organisms, they do not break down naturally and tend to accumulate in biota. Because of their persistence, bioaccumulation, potential for long-range environmental transport and toxicity, they are classified as persistent organic pollutants (POPs) and either listed or under consideration for listing in international agreements at regional or global level.

Although the Stockholm Convention on POPs is the latest of the international agreements, its global coverage and inclusion of all environmental media makes the Stockholm Convention the most important of these. However, in the other regional Conventions, a lot of experiences have been gained, which are used in the proposal to list SCCPs as new POPs in one of the annexes to the Stockholm Convention.

This chapter focuses on SCCPs with an emphasis on production, uses and international regulation.

2 Identity, Production, Toxicity, Fate and Transport of SCCPs

2.1 *Industrial Production*

At industrial scale, CPs are produced through chlorination of a petroleum-based hydrocarbon stream that has a distribution in carbon chain lengths. Individual chlorinated alkanes are typically not considered as CPs by the chemical industry. When a specific description is given for commercial CPs, it can be expected that the mixture will fall, on average, within that description, but other compounds may be present. Theoretically, SCCPs with the chemical formula $C_xH_{(2x-y+2)}Cl_y$, where $x = 10$–13 and $y \geq 1$, may contain a chlorine content ranging from 16% to 87% by weight. However, not all possible congeners would be produced in the industrial manufacturing process. For example, a product described as SCCP, 40% chlorine, will, on average, be composed of chlorinated alkanes that are 40% chlorine by weight and contain predominantly chain lengths between 10 and 13 carbons; the product may also contain lower and higher chlorinated alkanes as impurities. Currently, it is estimated that there are essentially no alkanes with <30% chlorination being produced and likely none with greater than 75% chlorination [3].

2.2 *SCCP Definitions and Their Relationship to CAS Numbers*

The numbers provided by the Chemical Abstracts Service (CAS) [4] are the unanimously accepted definitions for chemical substances. Given the complexity of this class of substances, an exhaustive listing of CAS numbers is not possible. Nevertheless, the following information contained in Table 1 provides a partial list of CAS numbers that contain chlorinated alkanes that fall within the defined range of SCCPs.

In addition, some CAS numbers denote individual chlorinated alkanes, not mixtures. Examples of these are in Table 2.

Table 1 CAS number and name of some chlorinated paraffin mixtures [3]

CAS No.	Substance name
51990-12-6	Chlorowax
61788-76-9	Alkanes, chloro
63449-39-8	Paraffin waxes and hydrocarbon waxes, chloro
68188-19-2	Paraffin waxes and hydrocarbon waxes, chloro, chlorosulphonated
68476-48-2	Hydrocarbons, C_2–C_6, chloro
68606-33-7	Hydrocarbons, C_1–C_6, chloro
68911-63-7	Alkanes, chloro, sulphurised
68920-70-7	Alkanes, C_6–C_{18}, chloro
68938-42-1	Paraffin waxes and hydrocarbon waxes, chloro, reaction products with naphthalene
68955-41-9	Alkanes, C_{10}–C_{18}, bromo chloro
68990-22-7	Alkanes,C_{11}–C_{14}, 2-chloro
71011-12-6	Alkanes, C_{12}–C_{13}, chloro
72854-22-9	Paraffin waxes and hydrocarbon waxes, chloro, sulphonated, ammonium salts
73138-78-0	Paraffins (petroleum), normal C_5–C_{20}, chlorosulphonated, ammonium salts
84082-38-2	Alkanes, C_{10}–C_{21}, chloro
84776-06-7	Alkanes, C_{10}–C_{32}, chloro
85422-92-0	Paraffin oils, chloro
85535-84-8	Alkanes (C_{10}–C_{13}), chloro (50–70%)
85535-85-9	Alkanes (C_{14}–C_{17}), chloro (40–52%)
85535-86-0	Alkanes (C_{18}–C_{28}), chloro (20–50%)
85536-22-7	Alkanes, C_{12}–C_{14}, chloro
85681-73-8	Alkanes, C_{10}–C_{14}, chloro
97553-43-0	Paraffins (petroleum), normal $C > 10$, chloro
97659-46-6	Alkanes, C_{10}–C_{26}, chloro
104948-36-9	Alkanes, C_{10}–C_{22}, chloro
106232-85-3	Alkanes, C_{18}–C_{20}, chloro
108171-26-2	Alkanes, C_{10}–C_{12}, chloro
108171-27-3	Alkanes, C_{22}–C_{26}, chloro

Table 2 CAS number and name of some individual chlorinated alkanes

CAS No.	Substance name
112-52-7	1-Chlorododecane
1002-69-3	1-Chlorodecane
2162-98-3	1,10-Dichlorodecane
3922-28-9	1,12-Dichlorododecane

2.3 Physical–Chemical Properties of SCCPs

Environmentally relevant physical–chemical properties for SCCPs are summarised in Table 3. The information is from Tomy et al. [5], Muir et al. [6], OSPAR [7] and the EU review of SCCPs [8].

Table 3 Environmentally relevant physical properties of SCCP congeners and mixtures of isomers

SCCP congener	% Cl	Vapor pressure (Pa)	Henry's law constant (Pa m^3/mol)	Water solubility (µg/L)	log K_{OW}[a]	log K_{OA}[a]	References[b]
$C_{10}H_{18}Cl_4$	50	0.028	17.7	328, 630, 2,370	5.93	8.2	1, 2, 3, 4
$C_{10}H_{17}Cl_5$	56	0.0040–0.0054	2.62–4.92	449–692	–	8.9–9.0	1, 2, 3
$C_{10}H_{16}Cl_6$	61	0.0011–0.0022	–	–	–	–	1
$C_{10}H_{13}Cl_9$	71	0.00024	–	400	5.64	–	4
$^{14}C_{11}$	59	–	–	150	–	–	5
$C_{11}H_{20}Cl_4$	48	0.01	6.32	575	5.93	8.5	1, 2, 3
$C_{11}H_{19}Cl_5$	54	0.001–0.002	0.68–1.46	546–962	6.20–6.40	9.6–9.8	1, 2, 3
$C_{11}H_{18}Cl_6$	58	0.00024–0.0005	–	–	6.40	–	1, 3
$C_{11.5}$	60	–	–	–	4.48–7.38	–	6
$^{14}C_{12}H_{21}Cl_5$	51	0.0016–0.0019	1.37	–	–	–	1
$C_{12}H_{20}Cl_6$	56	–	–	–	6.61	–	3
$^{14}C_{12}H_{20}Cl_6$	56	0.00014–0.00052	–	–	6.2	–	1, 7
$C_{12}H_{19}Cl_7$	59	–	–	–	7.00	–	3
$C_{12}H_{18}Cl_8$	63	–	–	–	7.00	–	3
$C_{12}H_{16}Cl_{10}$	67	–	–	–	6.6	–	7
$C_{13}H_{23}Cl_5$	49	0.00032	–	78	6.14	9.4	3
$C_{13}H_{22}Cl_6$	53	–	–	–	6.77–7.00	–	3
$C_{13}H_{21}Cl_7$	58	–	–	–	7.14	–	3
$C_{13}H_{16}Cl_{12}$	70	2.8×10^{-7}	–	6.4	7.207	–	4
C_{10}–C_{13}	49	–	–	–	4.39–6.93	–	6
C_{10}–C_{13}	63	–	–	–	5.47–7.30	–	6
C_{10}–C_{13}	70	–	–	–	5.68–8.69	–	6
C_{10}–C_{13}	71	–	–	–	5.37–8.69	–	6

[a]Octanol–air partition coefficient calculated from K_{OW}/K_{AW}, where K_{OW} is the octanol–water partition coefficient and K_{AW} is the air–water partition coefficient or unitless Henry's law constant (K_{AW} = HLC/RT, where HLC = Henry's law constant, R = gas constant 8.319 Pa m^3/mol K^{-1} and T = 293 K)

[b]References (cited in [3]): 1. Drouillard et al. (1998a), measured data; 2. Drouillard et al. (1998b), estimated data; 3. Sijm and Sinnige (1995), measured data; 4. BUA (1992), estimated data; 5. Madeley et al. (1983a), measured data; 6. Renberg et al. (1980), thin-layer chromatography (TLC) – K_{OW} correlation; 7. Fisk et al. (1998a), measured data

2.4 Production, Uses and Releases

In 2007, a survey on production and uses of SCCPs was undertaken in the framework of the United Nations Economic Comission for Europe (UNECE) POPs Protocol [9]. Based on the available data from the EU, Switzerland, Canada and the USA, production of SCCPs in the UNECE region was estimated to range from 7,500 tons per year to 11,300 tons per year (Table 4). According to the EU Risk Assessment Report, in 1994 SCCPs were manufactured within the EU by two producers, at a total quantity of <15,000 tons per year [8]. According to the updated draft Risk Assessment Report from August 2005 [10], SCCPs were produced by

Table 4 Estimated production of SCCPs in Europe, Canada and USA (data compiled in 2007 [9])

Region	Production (tons per year)	Year
EU (25)	1,500–2,500	2006
USA	6,000 and 8,800	2002/1998 and 2005
Canada	0	2006
Total	7,500–11,300	

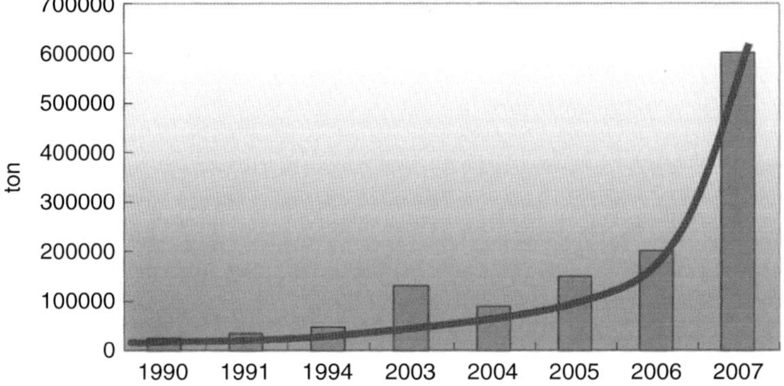

Fig. 1 The production of chlorinated paraffins in China in recent years

two companies, that is INEOS Chlor Ltd, UK (trade name: CERECLOR) and Caffaro Chimica S.r.l., Italy (trade name: Cloparin).

SCCPs were produced in Slovakia. Within the last years, a decrease in production volume could be observed: while in 2004, 560 tons were produced per year, the volume decreased to 354 tons per year in 2005 and 380 tons per year in 2006. For 2007, further decrease of production is anticipated in Slovakia. According to national answers to the UNECE questionnaire, production had stopped in Germany in 1995 and in Belgium. Switzerland never produced SCCPs [9].

The volume of SCCP produced in Russia, Taiwan and Brazil is unknown. 67,727 tons per year in metal works in eastern European UNECE countries is taken into account; however, additional production in the UNECE region may exist.

China began its CPs production at the end of 1950s. In 1980s, the production of CPs increased rapidly due to the high demand from the plastics industry. Whereas in the early 1980s, annual production of CPs was only a few thousand tonnes in China, the production increased to about 600,000 tonnes in 2007. It seems that currently, China is the largest producer in the world and the number of CP factories in China is more than 140. The time trend of CP production in China is shown in Fig. 1 [11–13].

In Canada in 2003 [14], and in the European Union (EU) in 1994 [15] and 1998 [7], the major uses and releases of SCCPs were in metalworking applications. In the

Table 5 Applications of SCCPs in Europe [3]

Application	1994 Data[a]		1998 Data[b]	
	tons/year	% of total use	tons/ year	% of total use
Metalworking lubricants	9,380	71.02	2,018	49.5
PVC plasticisers	Note 3	Note 3	13	0.3
Paints, adhesives and sealants	1,845	13.97	713	17.5
Leather fat liquors	390	2.95	45	1.1
Rubber/flame retardants/textiles/polymers (other than PVC)[c]	1,493	11.31	638	15.7
Other	100	0.75	648	15.9
Total	13,208	100	4,075	100

[a]Data from [13]
[b]Data from [7] Western Europe
[c]The given data did not include information specifically on usage in PVC

EU, 9,380 tons per year were used for metalworking in 1994. These amounts were reduced significantly in 1998 (2,018 tons per year). Other uses include paints, adhesives and sealants, leather fat liquors, plastics and rubber, flame retardants and textiles and polymeric materials (Table 6). The amounts of SCCPs used in the EU were reduced from 13,208 tons per year to 4,075 tons per year for all uses in 1994 and 1998, respectively. Since 2002, the use of SCCPs in the EU in metalworking and fat liquoring of leathers has been subject to restrictions under EU Directive 2002/45/EC.

The Review Committee of the Stockholm Convention compiled the applications of SCCPs in the Europe as shown in Table 5.

In 1994, 70 tons of SCCPs were used in Switzerland, and it is estimated that uses have reduced by 80%. The most widespread use of SCCPs in Switzerland was in joint sealants. In Germany, the most important uses (74% of the total) of SCCPs were banned by the EU directive 2002/45/EC. SCCPs have been used as a polychlorinated biphenyl (PCB) substitute in gaskets (e.g. splices, in buildings), and this may be a source when buildings are renovated. Brazil indicates that 300 tons per year is used in Brazil for the purposes of flame retardant in rubber, car carpet and accessories. Use of SCCPs in Australia decreased by 80% between 1998/2000 and 2002 to approximately 25 tons per year of SCCPs in the metalworking industry [16]. In the Republic of Korea and in 2006, SCCPs were mainly used in lubricant and additive agents. The release pattern and quantitative data are not available. Also, the Republic of Mauritius reported that it does use SCCPs [17].

Table 6 presents the most common uses and releases of SCCPs. When data on SCCPs were not available, data on CPs of no specified chain length is presented. There is currently no evidence of any significant natural source of CPs [18]. Anthropogenic releases of SCCPs into the environment may occur during production, storage, transportation, industrial and consumer usage of SCCP containing products, disposal and burning of waste and landfilling of products (Table 6). The possible sources of releases to water from manufacturing include spills, facility wash-down and storm water runoff. SCCPs in metalworking/metal cutting fluids

Table 6 Uses and Releases of SCCPs or CPs (various chain lengths)

Application/ use	% By weight of final product	Types of releases	Amounts released	References
Metalworking lubricants		Loss at production/ formulation site	Controlled CPs losses of 1–2%; 0.06 g/kg CPs consumed; loss of 23 tons SCCPs /year in mid-1990s in Europe; default emission factors for CPs are 0.005% to air and 0.25% to wastewater before on-site treatment	EC (2000); KEMI (1991); EU (2003)
		Loss from use	Carry-off from work pieces is 2.5 kg/site/year for small user (100-L capacity) and 2,500 kg/site/year for larger user (95,000-L capacity); annual losses of CPs from cutting fluid are 48%, 75% and 100% for large, medium and small machine shops, respectively; 18% loss of SCCPs to wastewater (733 tons/year in 1998 in the EU) and 3% disposed in landfill from use in metalworking fluids; 10% discharged to wastewater from use in water-based metalworking fluids; loss of CPs are 18.5% and 31.6% for oil-based and water-based metalworking fluids, respectively. Default emission factors for CPs are 0.02% to air for both types of fluids.	Government of Canada (1993a); EC (2000); UK Environment Agency (2003a)
Paints, adhesives and sealants	5–15% CPs (paints)	Loss at production/ formulation site	Insignificant (paint); Low or zero (sealants); 5% solid waste (sealants)	Zitko and Arsenault (1974); UK Environment Agency (2003b)
	10–15% CPs (typically for sealant)	Loss from use/ application	Waste during application may be disposed in landfill sites; default emission factors for thermosetting resins are 0% to air and 0.1% MCCPs and LCCPs to wastewater Emission factor of 0.15%/year for MCCPs	BRE (1998); UK Environment Agency (2003a)

	Up to 20% CPs for some applications (sealant)	Loss from leaching		UK Environment Agency (2003a)
Leather fat liquors	1% SCCPs or less (EU only)			EU Directive 2002/45/EC
Plastics and rubber	10.1–16.8% CPs (conveyor belts)	Loss at production/ formulation site	Default emission factors for plastic additives are 0.1% to air and 0.05% CPs to wastewater; default emission factors for thermosetting resins are 0% to air and 0.05% CPs to wastewater	BRMA (2001); UK Environment Agency (2001); BRE (1998)
	6.5% CPs (shoe soles) 13% CPs (Industrial sheeting)	Loss from use Losses from volatilisation	Loss during wear and abrasion of products 0.05% during lifetime of product	
Flame retardants	1–4% CPs (typically) Up to 15% CPs for some applications 1–10% SCCPs added to rubber			Zitko and Arsenault (1974); BUA (1992)
Textiles and polymeric materials		Loss at production/ formulation site	17% of SCCPs use in 1998	EC (2000)

References can be found in [3]

may also be released into aquatic environments from drum disposal, carry-off and spent bath use [19]. These releases are collected in sewer systems and ultimately end up in the effluents of sewage treatment plants. Information on percentage releases to sewage treatment plants or on removal efficiency is not currently available.

Other releases could include use of gear oil packages, fluids used in hard rock mining and equipment use in other types of mining, fluids and equipment used in oil and gas exploration, manufacture of seamless pipe, metalworking and operation of turbines on ships [20].

Landfilling is a major disposal route for polymeric products in Canada. CPs would be expected to remain stabilised in these products, with minor losses to wash off from percolating water. Leaching from landfill sites is likely to be negligible owing to strong binding of CPs to soils. Minor emissions of these products, which are effectively dissolved in polymers, could occur for centuries after disposal [21].

Polymer-incorporated CPs could also be released during recycling of plastics, which may involve processes such as chopping, grinding and washing. If released as dust from these operations, the CPs would be adsorbed to particles because of high sorption and octanol–air partition coefficients.

Table 7 presents scenarios for exposure of humans to SCCPs as has been compiled for the discussion on listing SCCPs as persistent organic pollutants in the Stockholm Convention. The first two scenarios assume consumption of

Table 7 Scenarios for exposure of humans and relevant toxicity values [3]

Receptor (kg)	Exposure (µg/kg bw/d)	Sample	Relevant toxicity value (µg/kg bw/d)	Comments
Adult	0.15–0.37	Estimated dietary exposure Baffin Island male Inuit	125,000	LOEL, 2 year rat study (NTP, 1986)
Adult	0.15–0.37	Estimated dietary exposure Baffin Island male Inuit	11	The 11 µg/kg bw/d was derived using a safety factor of 1,000 applied to a value of 11 mg/kg body weight, based on multistage modeling of tumors with the highest incidence in the carcinogenesis bioassay of male mice that resulted in 5% increase in tumor incidence [1, 21]
Breast fed child	0.0585	Human milk: 13 µg/kg lipid wt	125,000	LOEL, 2 year rat study (NTP, 1986)
Breast fed child	0.0585	Human milk: 13 µg/kg lipid wt	11	The 11 µg/kg bw/d was derived using a safety factor of 1,000 applied to a value of 11 mg/kg body weight, based on multistage modeling of tumors with the highest incidence in the carcinogenesis bioassay of male mice that resulted in 5% increase in tumor incidence [1, 21]

Table 8 Possible alternatives or substitutes to SCCPs taken from HELCOM and OSPAR [27], published in [9]

Use	Possible alternative to SCCP
Extreme pressure additive in metal working fluids	MCCPs, LCCPs, alkyl phosphate ethers, sulphonated fatty acid esters
Plastisers in paints	MCCPs
Additives in sealants	MCCPs, LCCPs, phthalate esters
Leather industry	LCCPs, natural animal and vegetable oils
Paints, coatings	LCCPs, phthalate esters, diisobutyrate and phosphate and boron containing compounds
Flame retardant in rubber, textiles and PVC	Antimony trioxide, aluminum trioxide, acrylic polymers and phosphate containing compounds
Rubber	LCCPs

traditional food diet. Exposure was calculated using SCCP concentrations measured in the Arctic in ringed seal blubber, beluga whale blubber and walrus blubber (from [22, 23]), and using dietary intake from Kuhnlein [24] and Kuhnlein et al. [25]. The scenarios for a breast-fed child assumes intake of 750 mL milk per day and 3% lipid content [26].

In addition, elevated levels of SCCPs in human breast milk in remote communities have been reported.

Information about substitutes for SCCPs is scarce. Some countries that responded to the UNECE survey in 2007 (Belgium, Czech Republic, Cyprus, Germany, the Netherlands, Canada and the USA) indicated to have no information on possible substitutes of SCCPs; France and Italy did not respond to that question. Switzerland and the United Kingdom responded that SCCPs in joint sealants, in leather processing and in paints and coatings were substituted by MCCPs (or a range of animal, vegetable and mineral oils). Euro Chlor understands from discussions with end-users that it would be difficult to substitute SCCPs, particularly in those applications where they are used as a flame retardant additive. In some cases, transition to MCCPs from SCCPs has already been possible; however, in some applications (e.g. as a flame retardant additive for rubber formulations), SCCPs enable the optimisation of high levels of chlorine content for maximum fire retardancy combined with workability of the formulation. An overview on possible alternatives or substitutes to SCCPs taken from HELCOM and OSPAR reports has been provided in TNO 2006 and is compiled in Table 8.

3 Status of SCCP under International Agreements

SCCP are addressed by various jurisdictions, although the scope of those reviews or actions has not necessarily always been the same. This chapter provides the definitions of SCCPs and present status of regulation as described in some international and national initiatives.

3.1 SCCPs under Regional Conventions

Table 9 provides an overview on the definitions of SCCP as described in some
regional international initiatives; this table had been submitted to initiate the
process for listing SCCPs under the Stockholm Convention [3]. It is noted that
there is no uniform nomenclature for SCCP, which may lead to differences in
interpretation of substance identity and possible ambiguity. For example, both the

Table 9 SCCP definition used in various assessments, legislation and/or jurisdictions

Jurisdiction/ overseeing body	Regulation/directive/ decision	Definition of SCCP
UNECE POPs Protocol		
OSPAR Commission	PARCOM Decision 95/1	CPs with carbon chain lengths between and including 10 and 13 and with a chlorination degree of more than 48% by weight
HELCOM		
European Commission	Recommendation 199/721/EC	$C_xH_{(2x-y+2)}Cl_y$, where $x = 10$–13 and $y = 1$–13 CAS No 85535-84-8, EINECS No 287-476-5
European Commission	Directive 2000/60/EC	C_{10}–C_{13} –chloroalkanes CAS No 85535-84-8, EINECS No 287-476-5
European Commission	Directive 2002/45/EC	Alkanes, C_{10}–C_{13}, chloro (SCCPs)
European Union	European Union Risk Assessment Report, Alkanes, C_{10}–C_{13}, Chloro-, CAS-No.: 85535-84-8, EINECS-No.: 287-476-5 [8]	Alkanes, C_{10}–C_{13}, chloro $C_xH_{(2x-y+2)}Cl_y$, where $x = 10$–13 and $y = 1$–13 CAS No. 85535-84-8, EINECS No. 287-476-5 Synonyms: alkanes, chlorinated; alkanes (C_{10}–C_{13}), chloro-(50–70%); alkanes (C_{10}–C_{12}), chloro-(60%); chlorinated alkanes, CPs; chloroalkanes; chlorocarbons; polychlorinated alkanes; paraffins-chlorinated Notes that there is a range of commercially available C_{10}–C_{13} CPs
European Commission	Risk Profile and Summary Report for SCCPs [10]	SCCPs are *n*-paraffins that have a carbon chain length of between 10 and 13 carbon atoms and a degree of chlorination of more than 48% by weight. There is a range of commercially available C_{10}–C_{13} CPs and they are usually mixtures of different carbon chain lengths and different degrees of chlorination although all have a common structure in that no secondary carbon atom carries more than one chlorine Alkanes, C_{10}–C_{13}, chloro $C_xH_{(2x-y+2)}Cl_y$, where $x = 10$–13 and $y = 1$–13 CAS No. 85535-84-8, EINECS No. 287-476-5 Synonyms: alkanes, chlorinated; alkanes (C_{10}–C_{13}), chloro- (50–70%); alkanes (C_{10}–C_{12}), chloro- (60%); chlorinated alkanes, CPs; chloroalkanes; chlorocarbons; polychlorinated alkanes; paraffins-chlorinated

Stockholm Convention Nomination (UNEP/POPS/POPRC.2/INF/6) and EC [10] define SCCP as C_{10}–C_{13}, >48% chlorine by weight and by the chemical formula $C_xH_{(2x-y+2)}Cl_y$, where $x = 10$–13 and $y = 1$–13, which translates to about 16–78% chlorine by weight (i.e. some compounds with <48% chlorine by weight).

3.1.1 UNECE LRTAP Convention: POPs Protocol

Since 1979, the Geneva Convention on Long-range Transboundary Air Pollution (LRTAP) has addressed some of the major environmental problems of the UNECE region through scientific collaboration and policy negotiation. The Convention has been extended by eight protocols that identify specific measures to be taken by Parties to cut their emissions of air pollutants. The aim of the Convention is that Parties shall endeavor to limit and, as far as possible, gradually reduce and prevent air pollution including long-range transboundary air pollution. Parties develop policies and strategies to combat the discharge of air pollutants through exchanges of information, consultation, research and monitoring [28].

In 1998, the Aarhus Protocol on POPs, which entered into force on 23 October 2003, presently has 29 Parties (February 2010). The Protocol's ultimate objective is to eliminate any discharges, emissions and losses of POPs. The Protocol bans production and use of some products right away; other POPs are scheduled for elimination at a later stage or severely restricted in use. The Protocol obliges Parties to reduce their emissions of polychlorinated-*p*-dioxins (PCDD), polychlorinated dibenzofurans (PCDF), polycyclic aromatic hydrocarbons (PAH), and hexachlorobenzene (HCB) below their levels in 1990 (or an alternative year between 1985 and 1995) [29].

In August, 2005, the European Community proposed SCCPs to be added to the UNECE LRTAP Convention, Aarhus Protocol on POPs. SCCPs were proposed to meet the criteria of decision 1998/2 of the Executive Body for persistence, potential to cause adverse effects, bioaccumulation and potential for long range transport. At the 24th session of the Executive Body in December 2006, the Parties to the UNECE POPs Protocol agreed that SCCPs should be considered as a POP as defined under the Protocol, and requested that the Task Force continue with the Track B reviews of the substances and explore management strategies for them. Finally, on 18 December 2009, the Executive Body for the UNECE LRTAP Convention at its 27th session agreed on revisions to the Protocol on POPs. The Parties broadened the Protocol's scope to include SCCPs together with six other POPs (hexachlorobutadiene, octabromodiphenyl ether, pentachlorobenzene, pentabromodiphenyl ether, perfluorooctane sulphonates and polychlorinated naphthalenes).

The following text was added to amend annex I to the POPs Protocol:

Substance	Implementation requirements	
	Elimination of	Conditions
SCCPs	Production	None, except for production for the uses specified in Annex II
	Use	None, except for the uses specified in Annex II

Annex II was amended as follows:

Substance	Substance requirements	
	Restricted to uses	Conditions
SCCPs	(a) Fire retardants in rubber used in conveyor belts in the mining industry	Parties should take action to eliminate these uses once suitable alternatives are available
	(b) Fire retardants in dam sealants	No later than 2015 and every 4 years thereafter, each Party that uses these substances shall report on progress made to eliminate them and submit information on such progress to the Executive Body. Based on these reports, these restricted uses shall be reassessed

In accordance with Article 14(3) of the POPs Protocol, the provisions in the amendment enter into force on the 90th day after the date on which two-thirds of the Parties to the POPs Protocol have deposited their instruments of acceptance thereof. After the entry into force of this Amendment, it shall enter into force for any other Party to the Protocol on the 90th day following the date of deposit of its instrument of acceptance.

3.1.2 OSPAR Commission

The OSPAR Convention for the Protection of Marine Environment of the North-East Atlantic is the current legal instrument guiding international cooperation on the protection of the marine environment of the North-East Atlantic. Work under the Convention is managed by the OSPAR Commission, made up of representatives of the Governments of 15 Contracting Parties[1] and the European Commission, representing the European Community [30].

In 1995, the OSPAR Commission adopted a decision on SCCPs (Decision 95/1) [31]. PARCOM Decision 95/1 requires Contracting Parties to phase-out the use of SCCPs as plasticiser in paints, coatings and sealants; as flame retardants in rubber, plastics and textiles; and their use in metalworking fluids. The phase-out for these uses should be achieved by 31 December 1999, except for uses as plasticiser in sealants in dams and as flame retardant in conveyor belts for the exclusive use in underground mining, which should be phased-out by 31 December 2004. The Recommendation initiates further study of other uses of SCCPs, which lead to diffuse discharges to the aquatic environment and the exchange of information on acceptable substitutes for SCCPs.

---

[1] It has been ratified by Belgium, Denmark, Finland, France, Germany, Iceland, Ireland, Luxembourg, Netherlands, Norway, Portugal, Sweden, Switzerland and the United Kingdom and approved by the European Community and Spain.

In 2006, OSPAR prepared an overview assessment of the implementation of PARCOM Decision 95/1 on SCCPs [32]. The assessment was based on national implementation reports received from nine of 15 Contracting Parties that have been requested to submit, in the 2005/2006 meeting cycle, reports on the national measures taken. All reporting Contracting Parties have taken measures to implement PARCOM Decision 95/1. Some Contracting Parties reported a full ban of all or certain uses of SCCPs and reductions of other uses. In general, Contracting Party measures have addressed those uses covered by Directive 2002/45/EC. A review of all remaining uses of SCCPs (not covered by Directive 2002/45/EC) will be carried out by the European Commission, in cooperation with the Member States and the OSPAR Commission, in the light of any relevant new scientific data on risks posed by SCCPs to health and the environment. The future EC risk reduction measures for the use of MCCPs may also be of relevance for the PARCOM Decision 95/1. Monitoring data on SCCPs will become available from reporting under the WFD, which lists SCCPs as priority hazardous substance, and could be used by OSPAR for further surveillance of the implementation of PARCOM Decision 95/1.

OSPAR 2006 agreed that further implementation reporting on PARCOM Decision 95/1 could cease for all Contracting Parties.

It should be noted that eleven of the EU Member States, which have fully endorsed the Commission Recommendation for SCCP (see Sect. 3.1.5), had committed themselves to PARCOM Decision 95/1 in 1995 under the Paris Convention (from 1998 OSPAR Convention). Thus, the PARCOM Decision goes further than the Commission Recommendation by including the uses of SCCPs as plasticisers and flame retardants. On the other hand, the Commission Recommendation goes further than the PARCOM Decision by including the uses of SCCPs in leather finishing. The European Community is not party to the PARCOM Decision. The United Kingdom did not accept the PARCOM Decision. Austria, Greece and Italy are not party to the OSPAR Convention.

3.1.3 HELCOM

The Helsinki Commission, or HELCOM, works to protect the marine environment of the Baltic Sea from all sources of pollution through inter-governmental co-operation between Denmark, Estonia, the European Community, Finland, Germany, Latvia, Lithuania, Poland, Russia and Sweden [33].

HELCOM has been assessing the effects of nutrients and hazardous substances on ecosystems in the Baltic Sea for the past over 25 years. Contaminants that are ecologically harmful are also referred to as pollutants or hazardous substances. In the Baltic Sea, these hazardous substances include the following [34]:

- Substances that do not occur naturally in the environment, such as PCBs, DDTs, dioxins, TBT, nonylphenolethoxylates (NP/NPE), SCCPs, brominated flame retardants and certain nitromusks
- Substances occurring at concentrations exceeding natural levels, including heavy metals like lead, copper, cadmium and mercury

HELCOM's objective with regard to hazardous substances is to prevent pollution of the convention area by continuously reducing discharges, emissions and losses of hazardous substances towards the target of their cessation by the year 2020, with the ultimate aim of achieving concentrations in the environment near background values for naturally occurring substances and close to zero for man-made synthetic substances. This objective was adopted within HELCOM Recommendation 19/5 in March 1998. The Recommendation contains a list of numerous substances of concern, from which HELCOM selected 42 hazardous substances for immediate priority action. These include biocides such as lindane and pentachlorophenol, metals and metal compounds such as mercury and lead, and industrial substances including SCCPs and nonylphenol.

To protect the marine environment of the Baltic Sea in a more holistic way, HELCOM adopted the Baltic Sea Action Plan (BSAP) in November 2007, introducing an ecosystem based approach to the management of human activities in the Baltic Sea region. With the adoption of the plan, HELCOM Contracting Parties committed themselves to, for example, work towards the goal to achieve a *Baltic Sea with life undisturbed by hazardous substances*. The goal is further defined by the following four ecological objectives:

- Concentrations of hazardous substances close to natural levels
- All fish safe to eat
- Healthy wildlife
- Radioactivity at pre-Chernobyl level

The action plan selected 11 hazardous substances/substance groups of priority concern, with corresponding indicators and targets, acknowledging the possible need for revision of the list and the actions in the future as more information becomes available. Amongst other actions agreed upon in the plan, Contracting Parties agreed to start to work for strict restrictions on the use of perfluorooctane sulphonate (PFOS), NP/NPEs and SCCPs in the whole Baltic Sea catchment area of the Contracting States. The SCCP-relevant entries are shown in Table 10.

The BSAP also reiterates HELCOM's commitment to contribute to work on hazardous substances in other international fora. HELCOM will provide coherent input from the Contracting States where possible based on a common HELCOM position, especially as regards the updating of lists of priority/candidate substances and substances to be evaluated, for example under the EU REACH, the 2001 Stockholm Convention on POPs and the 1998 Aarhus Protocol on POPs to the UNECE LRTAP Convention.

3.1.4 Mediterranean Action Plan and Barcelona Convention

As in the Baltic (see HELCOM), many human activities in the Mediterranean region result in environmental stresses and human health problems such as high nutrient concentrations and eutrophication, increasing incidences of algae blooms and increasing concentrations of hazardous substances. Regional multinational

Table 10 Chlorinated paraffins in the HELCOM list of substance relevant sectors of the 11 hazardous substances/substance groups of specific concern to the Baltic Sea

Substance	Main uses potentially relevant for the HELCOM area
8a. SCCPs (or chloroalkanes C_{10}–C_{13})	– Used in manufacture of textiles and wearing apparels in order to achieve clothes (designed, e.g., for sailing and industrial work) of high flame-resistant, water-proof and anti-fungal properties – Used as greasing agent in leather finishing, further use in manufacture of leather products – Used in metalworking fluids (both water- and oil-based) in treatment and coating of metal – Used as lubricants in compressed air tools in garages and in different industrial sectors – Used as plasticiser and flame retardant in paints (used, e.g., in road marking and as primer for surfaces exposed to sea water), varnishes and coatings – Used as plasticiser and flame retardant in rubber products such as gaskets, sealants and glues, which have been used, e.g., in construction sector and car industry – MCCP can contain up to 1% SCCP
8b. MCCPs (or chloroalkanes, C_{14}–C_{17})	– Used as substitute for SCCP – Used as greasing agent in leather finishing – Used in metalworking fluids (both water- and oil-based) in treatment and coating of metals – Used as plasticiser and flame retardant in paints (used, e.g., in road marking and as primer for surfaces exposed to sea water), varnishes and coatings – Used as plasticiser and flame retardant in rubber products such as gaskets and in glues which have been used, e.g., in construction sector and car industry – Used in some carbon copy paper types – Used as plasticiser and flame retardant in PVC plastic and further use in manufacture of plastic product

cooperation to abate these problems has mainly taken place under the umbrella of the 1975 Mediterranean Action Plan (MAP) and the 1976 Barcelona Convention (amended in 1995).

In the preparation for the global Stockholm Convention on POPs, A UNEP/GEF project was undertaken to characterise and prioritise POPs around the globe. To do so, the globe was divided into 12 regions, one of them being the Mediterranean. Within the Mediterranean RBAPTS project (regionally based assessment of POPs under UNEP/GEF in preparation for the Stockholm Convention on POPs), CPs (CAS No 85535848) were included into the list of 22 substances [35].

3.1.5 EU Water Framework Directive: Integrated River Basin Management for Europe

Directive 2000/60/EC of the European Parliament and of the Council of 23 October 2000 established a framework for Community action in the field of water policy.

It is addressed to the Member States of the European Union and entered into force on 22 December 2000 (the date of publication in the Official Journal (OJ L 327) [36]. The purpose of this directive is to establish a framework for the protection of inland surface waters, transitional waters, coastal waters and groundwater. This directive is to make a contribution towards enabling the Community and Member States to meet the international agreements containing important obligations on the protection of marine waters from pollution, in particular the Convention on the Protection of the Marine Environment of the Baltic Sea Area, signed in Helsinki on 9 April 1992 and approved by Council Decision 94/157/EC (HELCOM); the Convention for the Protection of the Marine Environment of the North-East Atlantic, signed in Paris on 22 September 1992 and approved by Council Decision 98/249/ EC (OSPAR); and the Convention for the Protection of the Mediterranean Sea Against Pollution, signed in Barcelona on 16 February 1976 and approved by Council Decision 77/585/EEC, and its Protocol for the Protection of the Mediterranean Sea Against Pollution from Land-Based Sources, signed in Athens on 17 May 1980 and approved by Council Decision 83/101/EEC (MAP). Further information on WFD can be found at http://ec.europa.eu/environment/water/water-framework/ implrep2007/index_en.htm.

Annex X of the final proposal for the EU-WFD contains a final list of 32 substances; among these are CPs (C_{10}–C_{13}) as determined by the European Council on 18 February 2000 in Brussels, Belgium.

The proposal for a directive of the European Parliament and of the Council amending for the 20th time Council Directive 76/769/EEC relating to restrictions on the marketing and use of certain dangerous substances and preparations (SCCPs) [37] contained the following:

In the EU, SCCPs are mainly used as an additive in metalworking fluids. Other uses are as a flame retardant in rubber formulations and as an additive in paints and other coating systems. Under Council Regulation (EEC) 793/93 on the evaluation and control of the risks of existing substances, the Commission's Draft Recommendation on the results of the risk evaluation and risk reduction strategies for four substances, including SCCP, was finalised in July 1998 and unanimously given a favorable opinion by the Article 15 Committee established under Regulation 793/ 93 on 28 July 1999. Following the approval by the College on 12 October 1999, the Recommendation was published in the Official Journal on 13 November 1999. This risk assessment concluded that there is a need for specific protective measures for the aquatic ecosystem.

The Commission Recommendation called for the consideration of measures at Community level, restricting the marketing and use of SCCP in particular in metalworking and leather finishing, adding that further work is necessary to establish those uses for which derogations can be justified. An independent study launched by DG III could not demonstrate evidence of a need for derogations for European industries.

Consulting the results of the Risk Assessment of SCCPs, the Scientific Committee on Toxicity, Ecotoxicity and the Environment (CSTEE) concluded on 27 November 1998 that there are potential unacceptable environmental risks associated with the life cycle of these chlorinated paraffins, though the use of SCCPs poses no significant risk to workers, consumers and man exposed via the environment.

On the basis of the Recommendation, and to avoid distortions of the internal market caused by the disparity of national legislation on CPs, the Commission proposes to introduce harmonisation measures in the framework of Directive 76/769 on restrictions of the marketing and use of certain dangerous substances and preparations. The 20th amendment of that Directive will ban SCCPs in the two areas of application cited by the Recommendation, namely in metalworking and leather finishing. With regard to the other applications of SCCPs, namely as plasticiser in paints, coatings and sealants, and as flame retardant in rubber, plastics and textiles, risk reduction measures should be reconsidered within three years of adoption of this Directive, in the light of the review of scientific knowledge and technical progress.

The following point is added to Annex I of Directive 76/769/EEC:

1. Alkanes, C_{10}–C_{13}, Chloro (SCCPs): CAS No. 85535-84-8 // 1. May not be placed on the market for use as substances and as constituents of preparations – in metalworking – for fat liquoring of leather.
2. Before 1 January 2003, the provisions on SCCPs will be reviewed by the European Commission in cooperation with the Member States in the light of any relevant new scientific data on risks to health and the environment of SCCPs.

The European producers of SCCPs offered a voluntary agreement to reduce risks to the marine environment by reducing production of SCCPs. Although this voluntary agreement was not accepted by the Commission, reductions have been achieved as can be seen in Table 11. The European producers have replaced SCCPs especially in the area of metalworking fluid by other additives, with the result that sales of SCCPs in the EU decreased from 13,000 tons in 1994 to 4,000 tons in 1998.

The Commission has to assess the progress in the implementation of the WFD in certain intervals and to inform the European Parliament, the Council and the public about the results of its assessments (see Article 18 WFD).

– First implementation report on the first stage of implementation (22 March 2007)
– Second implementation report on monitoring networks (1 April 2009)

The report of the European Commission on human health assessment in 2004 finds the following for C_{10}–C_{13} chloroalkanes [38]:

Table 11 SCCP sales per year in EU, specified by application in metric tons [8]

Application	1994	1998
PVC plasticisers		13
Metal working lubricants	9,380	2,018
Flame retardants textile and rubber	1,310	617
Waterproofing textile	183	21
Paints, sealants and adhesives	1,845	713
Leather fat liquors	390	45
Sales through distributors into above applications plus minor uses	100	648
Total	13,208	4,075

- *Environmental assessment*: The CSTEE notes that the WFD report concludes that biomagnification is relevant for the derivation of the QS for secondary poisoning. The CSTEE suggests that the proposed QS for secondary poisoning should be based on biota concentrations and the application of the more appropriate biomagnification models that we refer to in the Opinion on 'Marine TGD' and 'Chloroalkanes'.
- The CSTEE has produced an opinion on the RAR for this substance (Opinion on Alkanes, C_{10}–C_{13}, chloro {SCCP}, 6th CSTEE plenary meeting, 27 November 1998).
- *Human health assessment*: The CSTEE notes that, QS referring to food uptake by humans and drinking water abstraction, the WFD report is based on a recent EU risk assessment report. The conclusion that these quality standards are far higher than the standards needed to protect the aquatic community and are therefore not integrated into the definition of the is supported by the CSTEE.

3.2 Classifications and Some Conclusion

The SCCPs are according to a recent decision (in 25th Adaptation to Technical Progress of EU Directive 67/548/EEC on the classification, packaging and labeling of dangerous substances) classified as Dangerous for the Environment, with the symbol N and the risk phrases R50/53 (very toxic to aquatic organisms/may cause long-term adverse effects in the aquatic environment), and Harmful, Carcinogen, cat. 3 with the symbol Xn and risk phrase R40 (possible risk of irreversible effects) [39].

The Assessment report prepared by Sweden for HELCOM in June 2002 concluded other existing or new measures and instruments. There is no satisfactory overview of the status of CPs implementation of PARCOM Decision 95/1. In Finland and the Netherlands, national restrictions equivalent to the PARCOM Decision have been notified. In Norway, such a proposal is under consideration. In Sweden, a complete phasing-out of uses of SCCPs has taken place by voluntary means. Further, 90% of the use of MCCPs and LCCPs has been phased out. An almost complete phase-out of SCCPs used for the formulation of metalworking fluids seems to have taken place in Germany and Norway. Corresponding phasing-out activities are also reported from Belgium and UK. There is no information on phasing-out activities in remaining CPs.

3.3 Registers in the European Union

3.3.1 EU-EPER

The European Pollutant Emission Register (EPER) is the first European-wide register on the industrial releases into air and water. It was established through Decision 2000/479/EC. It provides data for large and medium-sized point sources in the industrial sectors covered by the IPPC Directive; however, it excludes

Table 12 Industrial emissions of C_{10}–C_{13} chloroalkanes to air and to water (direct and indirect emissions), respectively ([40], reference year is 2004)

Activity code	Activity description	Emission air (tons per year)	Emission water direct (tons per year)	Emission water indirect (transfer to an off-site waste water treatment) (tons per year)
4.1	Basic organic chemicals	–	–	0.01
1.1	Combustion installations >50 MW	–	–	0.00
6.7	Installations for surface treatment or products using organic solvents (>200 t/y)	–	–	0.00
5.1/5.2	Installations for the disposal or recovery of hazardous waste (>10 t/d) or municipal waste (>3 t/h)	–	0.01	–
Total (as reported in EPER database)		–	0.01	0.02

emissions from the transport sector and from most agricultural sources. According to EPER Decision, Commission Decision of 17 July 2000, EU Member States have to produce a triennial report, which covers the emissions of 50 pollutants from those industries whose emissions exceed established threshold values [40]. From the POPs, PCDD/PCDF and HCB are included in EPER.

According to EG-RL 96/61, C_{10}–C_{13} chloroalkanes with emissions greater than 1 kg per year into water have to be reported.

The summary in Table 12 shows the aggregated data from EPER (status December 2009) covering four activity levels and four installations in 25 EU Member States (EU25) for the reference year 2004. Two facilities are located in Italy and two in Spain. All of the direct emissions to water (100%) were reported from Italy, whereas only 9.5% indirect emissions to water were from Italy and the majority was from Spain.

3.3.2 EU-E-PRTR

Decision 166/2006/EC created the European Pollutant Release and Transfer Register (E-PRTR) [41]. The E-PRTR is the new Europe-wide register that provides easily accessible key environmental data from industrial facilities in European Union Member States and in Iceland, Liechtenstein and Norway. It replaces and improves the previous EPER.

The new register contains data reported annually by some 24,000 industrial facilities covering 65 economic activities across Europe.

For each facility, information is provided concerning the amounts of pollutant releases to air, water and land as well as off-site transfers of waste and of pollutants in wastewater from a list of 91 key pollutants including heavy metals, pesticides, greenhouse gases and dioxins for the year 2007. Some information on releases from diffuse sources is also available and will be gradually enhanced.

Table 13 Chloroalkanes pollutant releases reported in E-PRTR covering all E-PRTR states for reference year 2007

Releases per industrial activities	Facilities		Air	Water (kg)	Soil
Energy sector	1	Total	–	41.9	–
		Accidental	–	0	–
Chemical industry	1	Total	–	2.50	–
		Accidental	–	0	–
Waste and waste water management	8	Total	–	210	–
		Accidental	–	0	–
Paper and wood production processing	2	Total	–	36.2	–
		Accidental	–	0	–
Total	12	Total	–	290	–
		Accidental	–	0	–

Twelve facilities reporting; all values are annual releases
(*source*: http://prtr.ec.europa.eu/PollutantReleases.aspx)

Table 14 Chloroalkanes pollutant transfers reported in E-PRTR covering all E-PRTR states for reference year 2007

Transfer per industrial activities	Facilities	Quantity (kg)
Production and processing of metals	1	9.49
Chemical industry	2	12.8
Total	3	22.3

Three facilities reporting; all values are yearly transfers (http://prtr.ec.europa.eu/PollutantTransfers.aspx)

The register contributes to transparency and public participation in environmental decision-making. It implements for the European Community, the UNECE PRTR Protocol to the Aarhus Convention on Access to Information, Public Participation in Decision-making and Access to Justice in Environmental Matters.

The information for C_{10}–C_{13} chloroalkanes and the reference year 2007 on releases and transfers, respectively, provided in Tables 13 and 14 is the most recent information that could be accessed from the E-PRTR database.

3.4 SCCPs in National Legislation

Table 15 provides an overview on the definitions of SCCPs as described in some national initiatives. It is noted that there is no uniform nomenclature for SCCPs, which may lead to differences in interpretation of substance identity and possible ambiguity.

3.5 International Classifications of SCCPs

The International Agency for Research on Cancer (IARC) considers some SCCPs (average C_{12}, average 60% chlorination) to be possible carcinogens (group 2B),

Table 15 Definitions of SCCPs in some national initiatives

Country	Regulation/directive/decision	Definition of SCCP
Australia	SCCPs – Priority Existing Chemical Assessment Report No. 16 Ref.: NICNAS 2001	SCCPs contain between 10 and 13 carbon molecules. The assessment covers SCCPs generally; however, the following substances were specifically cited in the declaration notice: CAS No. 68920-70-7, Alkanes, C_6–C_{18}, chloro CAS No. 85535-84-8, Alkanes (C_{10}–C_{13}), chloro (50–70%) CAS No. 71011-12-6, Alkanes, C_{12}–C_{13}, chloro CAS No. 85536-22-7, Alkanes, C_{12}–C_{14}, chloro CAS No. 85681-73-8, Alkanes, C_{10}–C_{14}, chloro CAS No. 108171-26-2, Alkanes, C_{10}–C_{12}, chloro
Canada	Risk assessment (Follow-up Report for CPs) [42]	CPs are chlorinated derivatives of *n*-alkanes, having carbon chain lengths ranging from 10 to 38 and a chlorine content ranging from 30 to 70% by weight. SCCPs have carbon chains containing 10–13 carbon atoms
Japan	Chemical Substance Control Law Ref.: Japan submission via email	The following chemicals are listed in the existing chemical list: CPs (C = 20 to 32) Chlorinated normal paraffins (C = 8 to 22) Mono or dichloro alkanes (C = 6 to 24) No other definition, other than the above names, is given
Japan	vPvB regulation Ref.: Japan submission via email	The following chemical is listed: chlorinated paraffins (C = 11, Cl = 7–12) No other definition for CPs is given
Switzerland	Ordinance on Risk Reduction Related to the Use of Certain Particularly Dangerous Substances, Preparations and Articles (Ordinance on Risk Reduction related to Chemical Products (ORRChem)) Ref.: http://www.admin.ch/ch/f/rs/c814_81.html	From Annex 1.2: Paraffin chlorination products containing 10 to 13 carbon atoms (alkanes, C_{10}–C_{13}, chloro-) are SCCPs

(continued)

Table 15 (continued)

Country	Regulation/directive/decision	Definition of SCCP
United Kingdom	Updated Risk Assessment of Chloroalkanes, C_{10-13} Ref: UK Environment Agency 2007	Chlorinated *n*-paraffins with a carbon chain length of 10–13. It should be noted that around 40 CAS numbers have been used to describe the whole CP family at one time or another. The CAS number that is listed in IUCLID (85535-84-4) is taken to represent the commercial substance
United States	Ref: Toxics Release Inventory USEPA 1999	The US-EPA does not use the term "SCCPs" to define a category of chemicals subject to regulations. However, under the Toxics Release Inventory, the polychlorinated alkanes, C_{10}–C_{13}, category is defined as $C_xH_{(2x-y+2)}Cl_y$, where $x = 10$–13, $y = 3$–12, and the average chlorine content ranges from 40–70% with the limiting molecular formulas set at $C_{10}H_{19}Cl_3$ and $C_{13}H_{16}Cl_{12}$

although questions have been raised regarding the mechanisms for induction of tumors and the relevance for human health of the studies on which this classification was derived [43].

The current EU hazard classification for SCCPs is as follows: Carcinogen Category 3; R40 - N; R50-53 (Risk Phrases: R40: Limited evidence of a carcinogenic effect; R50/53: Very toxic to aquatic organisms; may cause long-term adverse effects in the aquatic environment.) ([8] in accordance with Directive 67/548/EEC).

The CSTEE suggests that the finding of lung tumors in male mice may be of importance for humans, but concluded in its risk characterisation that the use of SCCPs pose no significant risk for consumers or for man exposed via the environment [58]. The EU Risk Assessment Report [44] summarised the effects of SCCPs in mammalian species. Rodent studies showed dose-related increases in adenomas and carcinomas in the liver, thyroid and kidney. They concluded that there was insufficient evidence to conclude that the carcinogenicity observations in the liver and thyroid of mice and the benign tumors in the kidney of male rats were a male rat-specific event and, consequently, the concern for humans could not be ruled out. However, the EU risk assessment [8] also noted that, although there was an increase in alveolar/bronchiolar carcinomas in mice, the results were within historical control ranges, and the controls had a greater incidence of adenomas of the lung than the treated animals. The EU concluded that there was no significance for human health that could be read into this pattern of results.

According to the International Programme on Chemical Safety (IPCS), a tolerable daily intake (TDI) for SCCPs of 100 µg/kg b.w. per day is given [21].

The International Maritime Organisation (IMO) categorised CPs (C_{10}–C_{13}) as "Noxious liquid substances transported in bulk" (Appendix 2 of Annex to the 1973 Intervention Protocol) [45].

4 Stockholm Convention on Persistent Organic Pollutants

4.1 Brief Introduction and History of the Stockholm Convention on POPs

The Stockholm Convention on Persistent Organic Pollutants (POPs) is a global treaty to protect human health and the environment from POPs. POPs are chemicals that remain in the environment for long periods, become widely distributed geographically, accumulate in living organisms and are toxic to humans and wildlife. POPs may travel long distances through air, water or organisms and can cause damage wherever they travel. In implementing the Convention, governments will take measures to eliminate or reduce the releases of POPs into the environment through elimination of production and use of these chemicals or through reduction of releases from sources. The final goal of the convention is to eliminate intentionally produced and unintentionally generated POPs [46].

The text of the Stockholm Convention was negotiated and agreed in 2001 with a list of twelve initial POPs as shown below in the left column of Table 16. As can be seen, some of the POPs are single compounds, and others are mixtures of isomers or congeners. For three classes of POPs – PCBs, PCDD/PCDF and toxaphene – indicator congeners have been assigned to control enforcement of measures to eliminate POPs.

However, the convention is open for addition of new POPs to be included and to be listed in either of the Annexes A (elimination), B (permitted uses) or C (unintentional releases). To do so, a formal process has been set up in the convention; the process is led by the POP Review Committee (POPRC). The POPRC is a subsidiary body to the Stockholm Convention, established pursuant to paragraph 6(d) of Article 19 of the Convention. The mandate of the POPRC is to perform the functions assigned to it by the Convention, including the scientific review of the proposals and related information submitted by Parties to the Convention for listing new chemicals in Annex A, B and C according to Article 8 of the Convention, and to make recommendations to the Conference of the Parties.

4.2 The POPs Review Committee and the Process to List New POPs

The POPRC was set up by the Conference of the Parties to the Stockholm Convention at its first session in May 2005. Since then, the Committee has met five

Table 16 POPs chemicals presently recommended to be analyzed in the Global Monitoring Plan (GMP)

Chemical	Parent POPs	Transformation products
Aldrin	Aldrin	
Chlordane	*cis*- and *trans*-chlordane	*cis*- and *trans*-nonachlor, oxychlordane
Dichlorodiphenyltrichloro-ethane (DDT)	4,4′-DDT, 2,4′-DDT	4,4′-DDE, 2,4′-DDE, 4,4′-DDD, 2,4′-DDD
Dieldrin	Dieldrin	
Endrin	Endrin	
Hexachlorobenzene	HCB	
Heptachlor	Heptachlor	β-Heptachlorepoxide
Mirex	Mirex	
Polychlorinated biphenyls (PCB)	Sum of PCB$_7$ (7 congeners): 28, 52, 101, 118, 138, 153 and 180 PCB with TEFs[a] (12 congeners): 77, 81, 105, 114, 118, 123, 126, 156, 157, 167, 169 and 189	
Polychlorinated dibenzo-*p*-dioxins (PCDD) and polychlorinated dibenzofurans (PCDF)	2,3,7,8-Substituted PCD/PCDF (17 congeners with TEFs[b])	
Toxaphene	Congeners P26, P50, P62	

[a]PCB with toxic equivalency factors (TEFs) assigned by WHO [47]
[b]All 17 congeners with Cl substitution in the positions 2, 3, 7 and 8 have been assigned TEFs by WHO [47]

times – annually in the autumn of 2005, 2006, 2007, 2008 and 2009 – and accumulated a significant amount of experience in processing nominations for new chemicals to be listed under the Convention.

At its third and fourth meeting, the POPRC decided the listing of nine chemicals in Annexes A, B or C of the Convention, and to submit these recommendations to the Conference of the Parties for its consideration in accordance with paragraph 9 of Article 8 of the Convention. Subsequently, the conference of the parties at its fourth meeting in May 2009 adopted the recommendation of the POPRC and listed the nine chemicals as follows [48]:

(a) To list in Annex A of the Convention:

1. 2,2′,4,4′-Tetrabromodiphenyl ether (BDE-47, CAS No. 40088-47-9) and 2,2′,4,4′,5-pentabromodiphenyl ether (BDE-99, CAS No. 32534-81-9) and other tetra- and pentabromodiphenyl ethers present in commercial pentabromodiphenyl ether
2. Chlordecone
3. Hexabromobiphenyl
4. Lindane (gamma hexachlorocyclohexane, γ-HCH)
5. Alpha hexachlorocyclohexane
6. Beta hexachlorocyclohexane

   7. 2,2′,4,4′,5,5′-hexabromodiphenyl ether (BDE-153, CAS No. 68631-49-2), 2,2′,4,4′,5,6′-hexabromodiphenyl ether (BDE-154, CAS No. 207122-15-4), 2,2′,3,3′,4,5′,6-heptabromodiphenyl ether (BDE-175, CAS No. 446255-22-7) and 2,2′,3,4,4′,5′,6-heptabromodiphenyl ether (BDE-183 CAS No. 207122-16-5) and other hexa- and heptabromodiphenyl ethers present in commercial octabromodiphenyl ether

(b) To list in Annex B^2 of the Convention:
   8. Perfluorooctane sulphonic acid (CAS No. 1763-23-1), its salts and perfluorooctane sulphonyl fluoride (CAS No. 307-35-7)

(c) To list in Annex A and Annex C of the Convention:
   9. Pentachlorobenzene.

Although proposed by the European Commission in 2006, SCCPs were not yet recommended for listing in any of the annexes of the convention.

The POPs Review Process is a five-step process, which consists of the following steps:

1. *Propose a new chemical*: Any party may submit a proposal to the Secretariat for listing a new chemical, including information specified in Annex D. The Secretariat forwards the proposal to the POPRC.
2. *Apply screening criteria*: The POPRC examines the proposal and applies the *screening criteria* in Annex D.
3. *Develop a risk profile*: If the POPRC decides that the screening criteria have been fulfilled, the Secretariat invites all Parties and observers to provide technical comments and information specified in Annex E. The POPRC develops a *risk profile* based on the information.
4. *Develop a risk management evaluation*: If the POPRC decides on the basis of the risk profile that the proposal shall proceed, the Secretariat invites all Parties and observers to provide technical comments and socio-economic information specified in Annex F. The POPRC develops a *risk management evaluation* based on the information.
5. *List the chemical in Annex A, B and C*: The Conference of the Parties decides whether to list the chemical and specifies its related control measures in Annex A, B and C.

4.3   SCCPs under the Stockholm Convention

In May 2009, at the fourth meeting of the Conference of the Parties, nine candidate POPs have passed through all steps of the review process including adoption by the

---

^{2}The POPRC decision did not specify Annex B and recommended listing in either Annex A or Annex B.

Conference of the Parties and have become 'New POPs', so that the Stockholm Convention now has a total of 21 POPs.

Presently, there are three chemicals under review, namely (for SCCPs, the respective reference documents are provided, which can be accessed at www. pops.int) the following:

1. SCCPs
   - Proposal by EU to 2nd POPRC - Document UNEP/POPS/POPRC.2/14
   - Annex D: Screening - Decision POPRC.2/9
   - Annex E: Revised risk profile - UNEP/POPS/POPRC.4/10/Rev.1
   - Annex E: Supporting document - UNEP/POPS/POPRC.4/INF/20
2. Endosulfan (proposal by EU to 4th POPRC)
3. Hexabromocyclododecane (proposal by Norway to 4th POPRC)

4.4 The Proposal for Listing Short-Chained Chlorinated Paraffins in the Stockholm Convention on POPs

On 29 June 2006, the European Commission through its Directorate D: Water, Chemicals & Cohesion, D1, submitted to the Secretariat of the Stockholm Convention a letter requesting consideration of three candidate POPs for inclusion to the Stockholm Convention. These three chemicals submitted to the second meeting of the POPRC were the following:

- Octabromodiphenyl ether (CAS No.: 32536-52-0)
- Pentachlorobenzene (CAS No.: 608-93-5)
- SCCPs (alkanes, C_{10}–C_{13}, chloro)

While the proposals to list commercial octabromodiphenyl ether and pentachlorobenzene passed all steps of the review process quite smoothly and were listed as new POPs, the proposal for SCCPs presently stuck at step 2, the risk profile.

The information provided by the proponent and subsequent further information together with the evaluation of the Committee are presented in the next chapters. It should be noted that at present (December 2009) the POPRC was not able to reach consensus on adoption of the revised draft risk profile for SCCPs.

4.4.1 Chemical Identity of the Proposed Substance

The proposal nominated SCCPs (alkanes, C_{10}–C_{13}, chloro) with greater than 48% chlorination for listing as a POP under the Stockholm Convention. The proposal identified the substance as CAS No. 85535-84-8 and EINECS No. 287-476-5 (chloroalkanes, C_{10}–C_{13}). This CAS number represents the commercial SCCP product that is produced by the chlorination of a single hydrocarbon fraction consisting of n-alkanes that have a carbon chain length distribution consisting of

10, 11, 12 or 13 carbon atoms. As requested in Annex D, the following information on identity has been provided:

IUPAC Name	Alkanes, C_{10}–C_{13}, chloro
CAS No	85535-84-8
EINECS No	287-476-5
Synonyms	Chlorinated alkanes (C_{10}–C_{13})
	Chloro (50–70%) alkanes (C_{10}–C_{13})
	Chloro (60%) alkanes (C_{10}–C_{13})
	Chlorinated paraffins ($C_{10–13}$)
	Polychlorinated alkanes (C_{10}–C_{13})
	Paraffins chlorinated (C_{10}–C_{13})

SCCPs are chlorinated derivatives of *n*-alkanes having carbon chain lengths ranging from 10 to 13 and 1–13 chlorine atoms ($\sim$16–78% by weight, molecular formula $C_xH_{(2x-y+2)}Cl_y$, where $x = 10$–13 and $y = 1$–13). Chlorination of the *n*-alkane feedstock yields extremely complex mixtures, owing to the many possible positions for the chlorine atoms, and standard analytical methods do not permit their separation and identification. Thus the commercial mixture would fall under the proposed identity for SCCPs specified here.

The proposed Stockholm Convention nomination for listing is directed at SCCP products that contain more than 48% by weight chlorination. Figure 2 gives examples of two molecules within such SCCP product.

The POPRC has evaluated the SCCP proposal against the criteria listed in Annex D of the Stockholm Convention at the second meeting of the POPRC (Geneva, 6–10 November 2006). The Committee decided that SCCPs meet the screening criteria listed in Annex D of the convention (UNEP/POPS/POPRC.2/17 – Decision POPRC-2/8 Annex 1).

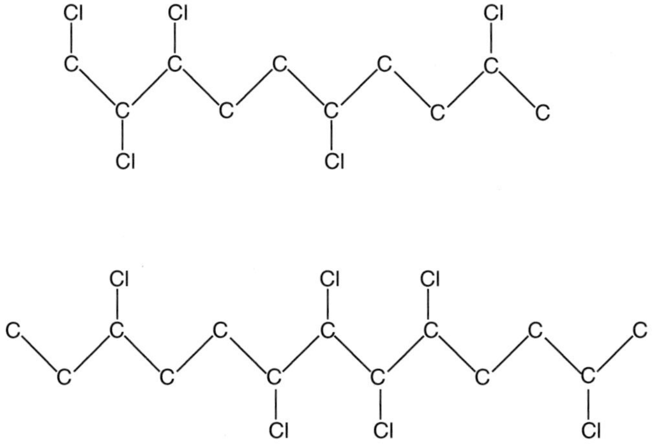

Fig. 2 Structure of two SCCP compounds ($C_{10}H_{17}Cl_5$ and $C_{13}H_{22}Cl_6$)

4.4.2 Synthesis of Information Gathered

SCCPs are persistent, bioaccumulative and toxic to some species, and undergo long-range transport to remote areas.

Total reported annual usage of SCCPs was high in several countries, but several have had notable reductions in recent years. For example, use in Canada was approximately 3,000 tons in 2000 and 2001 [49]; in Switzerland 70 tons were used in 1994, which is likely to be reduced by 80% now (Annex E submission); and in Western Europe, usage was reduced from 13,208 tons in 1994 [13] to 4,075 tons in 1998 [7]. As well, in Australia one of the two companies importing SCCPs in 1998/2000 had ceased importing by 2002. Furthermore, the use of SCCPs in Australia decreased by 80% during this period in the metalworking industry [50]. Releases can occur during production, storage, transportation and use of SCCPs. Facility wash down and spilled metalworking/metal cutting fluids are sources to aquatic ecosystems. Although data are limited, the major sources of release of SCCPs are likely the formulation and manufacturing of products containing SCCPs, such as PVC plastics, and use in metalworking fluids.

SCCPs are not expected to degrade significantly by hydrolysis in water, and dated sediment cores indicate that they persist in sediment more than 1 year. SCCPs have atmospheric half lives ranging from 0.81 days to 10.5 days, indicating that they are also relatively persistent in air. SCCPs also have vapour pressures in the range of known persistent organic pollutants that undergo long range atmospheric transport. In general, the HLC values reported imply that SCCPs' atmospheric transport is facilitated by remobilisation from water to air due to environmental partitioning. SCCPs have been detected in a diverse array of environmental samples (air, sediment, water, waster water, fish and marine mammals) and in remote areas such as the Arctic, which is additional evidence of long-range transport. In addition, Arctic Contamination Potential (ACP) modelling and the OECD LRTAP Screening Tool suggests that SCCPs have moderate ACP when emitted to air and have properties similar to known POPs that are known to undergo long-range transport.

Bioaccumulation factors (BAFs) of 16,440–25,650 wet weight (wet wt.) in trout from Lake Ontario indicate that SCCPs can bioaccumulate to a high degree in aquatic biota. This is supported by modelling data for log K_{ow} and bioaccumulation factors, which indicate that SCCPs bioaccumulate. In addition, biomagnification factors for some SCCPs have been found to be greater than 1. High concentrations of SCCPs in upper trophic-level organisms, notably in marine mammals and aquatic freshwater biota, are additional evidence of bioaccumulation.

SCCP has so far been found in arctic whales, seals, walruses [51] and two species of Arctic birds (Little auk, *Alle alle* and Kittiwake, *Rissa tridactyla*) [52]. Thus, SCCPs have been measured in animal species living in 'remote areas'. The following also shows that these concentrations are similar to concentrations of well-recognised POPs such as PCB, DDT and toxaphene [53]. More detailed comparisons of POP concentrations for the arctic marine mammals and for the bird Kittiwake (*Rissa tridactyla*) are presented in Table 17 and Table 18, respectively.

Table 17 Comparison by ratio of mean concentrations of SCCPs and POPs in Arctic species

Species	[sumPCB]/[SCCPs]	[sumDDT/SCCPs]	[Toxaphene/SCCPs]
Beluga whale	19–24	11–18	15
Ringed seals	2.3	1.3	0.9
Walrus	0.4	0.1	0.6

Mean SCCPs concentrations were 0.2 µg/g, 0.5 µg/g and 0.4 µg/g blubber in beluga whales, ringed seals and walrus, respectively [51, 53]

Table 18 Comparison of concentrations of SCCPs and POPs (ng/g lipid weight) in Arctic Kittiwake [52, 53]

Species	[SCCPs]	[sumDDT]	[sumPCB]
Kittiwake	110–880 ($n = 2$)	500–1,900	10,000–21,000

SCCPs have also been measured in the breast milk of Inuit women in Northern Quebec, as well as of women in the United Kingdom.

The hazard assessments for SCCPs and MCCPs have shown that these structurally very similar substances also have very similar hazard profiles. Both substances have a similar potency (i.e. NOAELs of the same order of magnitude) (see EU RARs on SCCPs and MCCPs); target organs in mammals include the liver, kidney and thyroid. Reth et al. [52] have measured MCCPs in two arctic bird species at concentrations somewhat exceeding the ones of SCCPs. As noted in Annex E, the risk profile can include 'consideration of toxicological interactions involving multiple chemicals', which in this case would be to consider the combined risk from exposure to both SCCPs and MCCPs. Therefore, the present risk profile could underestimate the risks from SCCPs in the presence of MCCPs.

There is evidence that SCCPs are toxic. The most sensitive organism, *Daphnia magna*, has a chronic NOEC of 5 µg/L. Japanese medaka was also very sensitive to SCCPs. The NOAEL was reported to be 9.6 µg/L.

Table 19 provides a Predicted environmental concentration (PEC), a critical toxicity value (CTV) and a Predicted no-effect concentration (PNEC) based on available empirical data for each class of receptors (e.g. pelagic organisms, benthic organisms). The maximum reported value was used as the PEC for each medium. A CTV typically represents the most sensitive chronic toxicity value and does not incorporate application factors ('safety factors') to account for uncertainties, nor for conservative approaches that could be considered for persistent and bioaccumulative substances. Some jurisdictions use NOECs to determine the CTV; however, given the paucity of this type of information for SCCP, LOECs were used instead. Application factors were applied to estimate PNECs. A value of 10 was used for extrapolating from a LOEC/LOAEL to a NOEC/NOAEL, and a value of 100 for laboratory to field and intra- and interspecies variations. Thus, a total value of 1,000 was used to derive the PNECs. It should be noted that risk may be underestimated using standard risk quotient methods because persistent chemicals may take a long time to reach maximum steady-state concentrations in

Table 19 List of estimated exposure values (EEV), critical toxicity values (CTV) and RQs for SCCPs

Receptor	EEV	Sample	CTV	PNEC	RQ
Pelagic	44.8 ng/L[a]	STP, Hamilton, ON	8,900 ng/L[b]	8.9	5
Pelagic	2.63 µg/g ww[c]	Lake Ontario	0.79 µg/g ww[d]	0.00079	3,329
Benthic	0.41 mg/kg[e]	Lake Ontario sediment	35.5 mg/kg[f]	0.0355	11.5
Soil dwelling	3.2 mg/kg[g]	UK sewage after 10 years soil application	1,230 mg/kg[h]	1.23	2.6
Microorganisms (bacterial, soil nitrification)	3.2 mg/kg[g]	UK sewage after 10 years soil application	570 mg/kg[i]	0.57	5.6
Mammals	2.63 mg/kg[c]	Carp from Hamilton Harbour, Lake Ontario	1,000 mg/kg food wet wt.[j]	1	2.63

[a]A dilution factor of 10 applied to the value for final effluent of sewage treatment plant in Hamilton Ontario (448 ng/L) (Environment Canada 2005) [42]
[b]Twenty-one-day chronic LOEC value for Daphnia magna [54]
[c]Measured SCCPs in carp collected in Lake Ontario in 1996 and 2001 [55, 56]
[d]Severe liver histopathologies; extensive fibrous lesions and hepatocyte necrosis of rainbow trout [57]
[e]Measured in surface sediments from Lake Ontario [58]
[f]Calculated using the LOEC for Daphnia magna using equilibrium partitioning approach [42]
[g]Estimated in Sect. 2.4.3 of [3]
[h]Experimental data for soil organisms and micoorganisms reported by Bezchlebova et al. [59]
[i]Soil nitrification [60]
[j]See Sect. 2.5.2.6 of [3] for calculation

environmental compartments and in tissues of laboratory organisms. Moreover, because food consumption is usually the primary route of exposure to POPs and persistent, bioaccumulative and toxic (PBT) substances in the field, PNECs may underestimate effect thresholds if the food pathway is not considered in key toxicity studies. Notwithstanding these uncertainties, risk quotients were used to illustrate potential risks. Risk quotients (RQ) were derived to estimate risk for the different classes of receptors.

These RQs show that all species could be at risk from exposure to SCCPs. In addition, elevated levels of SCCPs in human breast milk in remote communities have been reported.

The International Agency for Research on Cancer considers some SCCPs (average C_{12}, average 60% chlorination) to be possible carcinogens (groups 2B), although questions have been raised regarding the mechanisms for induction of tumors and the relevance for human health of the studies on which this classification was derived. The Science Committee on Toxicity, Ecotoxicity and the Environment suggests that the finding of lung tumors in male mice may be of importance for humans, but this information would not alter the conclusion of its risk characterisation that the use of short-chain chlorinated paraffins poses no significant risk for consumers or for man exposed via the environment [61]. The EU Risk Assessment Report [8] summarised the effect of SCCPs in mammalian species. Rodent studies showed dose-related increases in adenomas and carcinomas in the liver, thyroid and kidney. They concluded that there was insufficient evidence to conclude that the carcinogenicity observations in the liver and thyroid in mice and the benign tumors in the kidney of male rats were a male rat-specific event, and consequently the concern for humans could not be ruled out. However, the EU risk assessment [8] also noted that, although there was an increase in alveolar/bronchiolar carcinomas in mice, the results were within historical control ranges and the controls had a greater incidence of adenomas of the lung than the treated animals. The EU concluded that there was no significance for human health that could be read into this pattern of results. An independent technical peer review on SCCPs submitted under the UNECE-LRTAP POPs Protocol indicated that aboriginal people living in the Arctic and consuming contaminated animals may be exposed to SCCPs at concentrations greater than the WHO health guideline of 11 µg/kg bw for neoplastic effects (tumor formation) (UNECE-LRTAP POPs Protocol) [28]. A tolerable daily intake (TDI) for SCCPs of 100 µg/kg-bw per day is given by IPCS [21]. Although expert groups have different opinions as to the interpretation of these data, evidence of toxicity and exposure suggests that humans could be at risk.

SCCPs are persistent and bioaccumulative, and thus concentrations in the environment and biota are expected to increase with continued release to the environment. Standard risk assessment methods comparing effect levels to environmental concentrations may underestimate the risk of persistent and bioaccumulative substances, such as SCCPs. Persistent substances can take decades to reach a maximum steady state concentration in the environment, resulting in an underestimation of the potential exposure to these compounds if steady-state has not been achieved, and releases into the environment continue. Similarly, it can take a long time for persistent and bioaccumulative substances to reach a maximum steady-state concentration within an organism; this is supported by the observations of Sochová et al., [62] who noted an increase in toxicity of SCCPs for longer exposure duration with nematodes. The durations of standard toxicity tests may be insufficient to achieve the maximum tissue concentration, resulting in an underestimation of the effect threshold.

4.4.3 Concluding Statement

In summary, the increasing regulation of SCCPs has resulted in a decrease in SCCPs currently in use. However evidence suggests that significant amounts are still in use and being released in several countries. The available empirical and modeled data strongly indicate that SCCPs are persistent, bioaccumulative and toxic to aquatic organisms at low concentrations. In mammals, SCCP may affect the liver, the thyroid hormone system and the kidneys, for example, by causing hepatic enzyme induction and thyroid hyperactivity. SCCPs have characteristics similar to known POPs that undergo long-range environmental transport. SCCPs are considered as POPs pursuant to decisions taken under the UNECE LRTAP. Concentrations in biota and sediment from remote Arctic locations also suggest that long-range transport of SCCPs is occurring via air or ocean currents. SCCPs are present in Arctic marine mammals, which are in turn food for northern indigenous people. SCCPs are found in human breast milk both in temperate and Arctic populations. Simultaneous exposure to the analog MCCPs would increase the risks because of similar toxicity profiles of SCCPs and MCCPs.

Based on the available evidence, it is concluded that SCCPs are likely, as a result of their long-range environmental transport, to lead to significant adverse environmental and human health effects, such that global action is warranted.

4.4.4   Final Decision and Present Status (Presented at and After POPRC-5)

At its fourth meeting, the POPRC considered an updated draft risk profile on SCCPs [63] prepared by the ad hoc working group established in accordance with decision POPRC-3/8.

The Committee agreed to postpone further consideration of the draft risk profile to the fifth meeting of the Committee as it could not reach consensus on adoption of the draft risk profile. At the fifth meeting of the POPRC in October 2009 and despite submission of additional information such as the study by Japan on bioconcentration of CPs [64], the Committee was not in consensus, as it revealed that twelve members would favour and nine would oppose the proposal, while eight would abstain. To move proceed with the proposal, consensus would be necessary. Therefore, after POPRC-5, the proposal for listing SCCPs in either Annexes A, B or C of the Stockholm Convention remains in the Annex E phase (risk profile) for further consideration at POPRC-6 (which is scheduled for 18–22 October 2010 in Geneva). The Committee agreed with the provision that parties and observers would be invited to provide additional information on production data, inventories of uses, information on releases and additional information that could assist with evaluation, including on toxicity and ecotoxicity and on national and international risk evaluations.

References

1. EHC (1996) Environmental Health Criteria 181 – chlorinated paraffins. World Health Organisation. http://www.inchem.org/documents/ehc/ehc/ehc181.htm
2. UK Environment Agency. http://www.environment-agency.gov.uk/business/default.aspx
3. UNEP (2009) Supporting document for the risk profile on short-chained chlorinated paraffins. UNEP/POPS/POPRC.5/INF/18; for download http://chm.pops.int/Portals/0/download.aspx?d=UNEP-POPS-POPRC.5-INF-18.English.pdf
4. Chemical Abstracts Service, Colombus, OH, USA
5. Tomy GT, Fisk AT, Westmore JB, Muir DCG (1998) Environmental chemistry and toxicology of polychlorinated n-alkanes. Rev Environ Contam Toxicol 158:53–128
6. Muir DCG, Stern GA, Tomy GT (2000) Chlorinated paraffins. In: Paasivirta J (ed) New types of persistent halogenated compounds, vol 3, Part K, The Handbook of Environmental Chemistry. Springer, Berlin, pp 203–236

7. OSPAR (2001) OSPAR (Oslo-Paris convention for the protection of the marine environment of the North-East Atlantic). OSPAR draft background document on short chain chlorinated paraffins. 65 pp (OSPAR 01/4/8-E)

8. EC (2000) Opinion of the Economic and Social Committee on the proposal for a directive of the European parliament and of the council amending for the 20th time council directive 76/769/EEC relating to restrictions on the marketing and use of certain dangerous substances and preparations (short chain chlorinated paraffins). COM(2000) 260 final – 2000/0104 (COD)

9. UNECE (2007) Study contract on "support related to the international work on Persistent Organic Pollutants (POPs)" management option dossier for Short Chain Chlorinated Paraffins (SCCPs), 12 June 2007. Updated version on the basis of the outcome of the Sixth Meeting of the Task Force on POPs, 4-6 June 2007, Vienna, Austria, Service Contract ENV.D.1/SER/2006/0123r, DG Environment, European Commission. Report prepared by BiPRO. http://www.unece.org/env/lrtap/Task-Force/popsxg/2007/POST-Vienna%20Documents_%20revised%20AFTER%206th%20meeting/SCCP_Management_Options_070612-fin.doc

10. EC (2005) Risk profile and summary report for Short-Chained Chlorinated Paraffins (SCCPs). Dossier prepared from the UNECE convention on long-range transboundary air pollution, protocol on persistent organic pollutants. European Commission, DG Environment

11. Yuan B, Wang Y, Fu J, Jiang G (2009) Evaluation of the pollution levels of short chain chlorinated paraffins in soil collected from an e-waste dismantling area in China. Organohalogen Compd 71:3106–3109

12. Wang Y, Fu J, Jiang G (2009) The research of environmental pollutions and toxic effect of short chain chlorinated paraffins. Environ Chem 28:1–9 (in Chinese)

13. Yuan B, Wang Y, Fu J, Jiang G (2010) Study on the analytical method of chlorinated paraffins and the determination of chlorinated paraffins in soil samples. Chinese Science Bulletin (in press)

14. Environment Canada (2003) Data collected from "Notice with Respect to Short-, Medium- and Long-Chain Chlorinated Paraffins." Canada Gazette, Part I, November 30, 2002 (cited in [2])

15. Euro Chlor (1995) As reported in letter from ICI dated 12/7/95 (cited in [7])

16. NICNAS (2004) Environmental exposure assessment of short chain chlorinated paraffins (SCCPs) in Australia July, 2004. A follow up report to the National Industrial Chemicals Notification and Assessment Scheme (NICNAS) Short Chain Chlorinated Paraffins (SCCPs) priority existing chemical assessment report No. 16. Australian Government, Department of Health and Ageing, National Industrial Chemicals Notification and Assessment Scheme (NICNAS) (cited by [2])

17. Comments submitted to the Stockholm Convention Secretariat on 7 April 2008 for POPRC SCCPs risk profile

18. U.K. Environment Agency (2003) Updated risk assessment of alkanes, C_{10-13}, chloro (cited by [2])

19. Government of Canada (1993) Canadian Environmental Protection Act. Priority Substances List supporting document. Chlorinated paraffins. Environment Canada and Health and Welfare Canada. 66 pp (cited by [2])

20. CPIA (2002) Environment Canada (2003): Data collected from "Notice with Respect to Short-, Medium- and Long-chain Chlorinated Paraffins." Canada Gazette, Part I, November 30, 2002

21. IPCS (1996) Chlorinated paraffins. Environmental health criteria, vol 181. World Health Organization, Geneva, 181 pp

22. Tomy GT, Stern GA, Lockhart WL, Muir DCG (1999) Occurrence of C_{10}–C_{13} polychlorinated n-alkanes in Canadian mid-latitude and Arctic lake sediments. Environ Sci Technol 33:2858–2863

23. Muir DCG, Alaee M, Butt C, Braune B, Helm P, Mabury S, Tomy G, Wang X (2004) New contaminants in Arctic biota. Synopsis of research conducted under the 2003–2004, Northern contaminants program. Ottawa, Indian and Northern Affairs Canada 139–148

24. Kuhnlein HM (1995) Benefits and risks of traditional food for indigenous peoples: focus on dietary intakes of Arctic men. Can J Physiol Pharmacol 73:765–771
25. Kuhnlein HV, Receveur O, Muir DC, Chan HM, Soueida R (1995) Arctic indigenous women consume greater than acceptable levels of organochlorines. J Nutr 125:2501–10
26. Van Oostdam J, Gilman A, Dewailly É, Usher P, Wheatley B, Kuhnlein H, Neve S, Walker J, Tracy B, Feeley M, Jerome V, Kwavnick B (1999) Human health implications of environmental contaminants in Arctic Canada: a review. Sci Total Environ 230:1–82
27. TNO (2006) Van der Gon et al.: Study to the effectiveness of the UNECE persisitent organic pollutants (POP) protocol and costs of additional measures (Phase II: Estimated emission reduction and cost of options for a possible revision of the POP Protocol); July 2006, prepared for Netherlands Ministry of Housing, Spatial Planning and the Environment; 2006-A-R0187/B, order no. 35096 (cited by [3])
28. UNECE – United Nations Economic Commission for Europe. www.unece.org
29. UNECE (1998) Convention on Long-range Transboundary Air Pollution, the 1998 Aarhus Protocol on Persistent Organic Pollutants (POPs). http://www.unece.org/env/lrtap/pops_h1.htm
30. Oslo Paris Commission. www.ospar.org
31. PARCOM (1995) PARCOM Decision 95/1 on the phasing out of short chained chlorinated paraffins. Brussels, 26–30 June 1995. http://www.ospar.org/html_documents/ospar/html/ospar_decs_recs_other_agreements.pdf
32. OSPAR (2006) Hazardous substances series – overview assessment: implementation of PARCOM Decision 95/1 on short chained chlorinated paraffin. http://www.ospar.org/v_measures_spider/get_page.asp?v0=%2Fdocuments%2Fdbase%2Fdecrecs%2Fimplementation%2Fpd95%2D1%2Edoc
33. Helsinki Commission. www.helcom.fi
34. http://www.helcom.fi/environment2/hazsubs/en_GB/front/
35. UNEP (2002) Regionally based assessment of persistent toxic substances – Mediterranean regional report. United Nations Environment Programme, Geneva, December 2002. http://www.chem.unep.ch/pts/regreports/Mediterranean.pdf
36. http://eur-lex.europa.eu/LexUriServ/LexUriServ.do?uri=CELEX:02000L0060-20090113:EN:NOT
37. Proposal for a directive of the European Parliament and of the Council amending for the 20th time Council Directive 76/769/EEC relating to restrictions on the marketing and use of certain dangerous substances and preparations (Short Chain Chlorinated Paraffins) /* COM/2000/0260 final – COD 2000/0104 */ Official Journal C 337 E, 28/11/2000 P. 0138–0139
38. EC (2004) European Commission – Health & Consumer Protection Directorate-General, Directorate C – Public Health and Risk Assessment, C7 – Risk Assessment, Brussels, C7/GF/csteeop/WFD/280504 D(04): Opinion of the scientific committee on toxicity, ecotoxicity and the environment (CSTEE) on "The setting of Environmental Quality Standards for the Priority Substances included in Annex X of Directive 2000/60/EC in Accordance with Article 16 thereof " Adopted by the CSTEE during the 43rd plenary meeting of 28 May 2004 http://ec.europa.eu/health/ph_risk/committees/sct/documents/out230_en.pdf
39. HELCOM (2002) Helsinki Commission Baltic Marine Environment Protection Commission Implementing the HELCOM Objective with regard to Hazardous Substances "Implementation of the HELCOM Objective with regard to Hazardous Substances" – Project funded by European Communities (Subv 99/79391), Sweden and HELCOM. Guidance document on short chained chlorinated paraffins (SCCP) presented by Sweden, June 2002
40. EPER pollutants' inventories. http://eper.ec.europa.eu/
41. European Pollutant Release and Transfer Register (PRTR). http://prtr.ec.europa.eu/PollutantReleases.aspx
42. Environment Canada (2004) Follow-up report on PSL1 substance for which there was insufficient information to conclude whether the substance constitutes a danger to the environment; chlorinated paraffins. Existing Substances Division, Environment Canada, Gatineau, Quebec

43. IARC (1990) Summaries and evaluations chlorinated paraffins (Group 2B) Vol. 48. p 55
44. EU (2000) European Union risk assessment report. 1st Priority List Vol. 4: alkanes, C10–13, chloro-. European Chemicals Bureau, Luxembourg. 166 pp (EUR 19010; ISBN 92-828-8451-1)
45. IMO (1996) Resolution MEPC. 72 (38) adopted on 10 July 1996 – revision of the list of substances to be annexed to the protocol relating to the intervention on the high seas in cases of marine pollution by substances other than oil. MEPC 38/20, International Maritime Organisation
46. UNEP (2001) Stockholm Convention on Persistent Organic Pollutants. www.pops.int
47. van den Berg M, Birnbaum LS, Denison M, De Vito M, Farland W, Feeley M, Fiedler H, Hakansson H, Hanberg A, Haws L, Rose M, Safe S, Schrenk D, Tohyama C, Tritscher A, Tuomisto J, Tysklind M, Walker N, Peterson RE (2006) The 2005 World Health Organization re-evaluation of human and mammalian toxic equivalency factors for dioxins and dioxin-like compounds. Tox Sci 93:223–241
48. UNEP (2009) Report of the conference of the parties of the Stockholm Convention on Persistent Organic Pollutants on the work of its fourth meeting. UNEP/POPS/COP.4/38. www.pops.int. Accessed 8 May 2009
49. Environment Canada (2003) Data collected from "Notice with Respect to Short-, Medium- and Long-chain Chlorinated Paraffins." Canada Gazette, Part I, November 30, 2002
50. NICNAS (2004) Environmental exposure assessment of short chain chlorinated paraffins (SCCPs) in Australia July, 2004. A follow up report to the National Industrial Chemicals Notification and Assessment Scheme (NICNAS) Short chain chlorinated paraffins (SCCPs) priority existing chemical assessment report No. 16. Australian Government, Department of Health and Ageing, National Industrial Chemicals Notification and Assessment Scheme (NICNAS)
51. Tomy GT, Muir DCG, Stern GA, Westmore JB (2000) Levels of C_{10}–C_{13} polychloro-n-alkanes in marine mammals from the Arctic and the St Lawrence River estuary. Environ Sci Technol 34:1615–1619
52. Reth M, Ciric A, Christensen GN, Heimstad ES, Oehme M (2006) Short- and medium-chain chlorinated paraffins in biota from the European Arctic- differences in homologue group patterns. Sci Total Environ 367:252–260
53. AMAP (2004) AMAP assessment 2002: persistent organic pollutants in the Arctic. Arctic Monitoring and Assessment Programme (AMAP), Oslo, Norway. xvi + 310 pp
54. Thompson RS, Madeley JR (1983) The acute and chronic toxicity of a chlorinated paraffin to *Daphnia magna*. Imperial Chemical Industries PLC, Devon, U.K. (Brixham Report BL/B/2358)
55. Muir DCG, Bennie D, Teixeira C, Fisk AT, Tomy GT, Stern GA, Whittle M (2001) Short chain chlorinated paraffins: are they persistent and bioaccumulative? In: Lipnick R, Jansson B, Mackay D, Patreas M (eds) Persistent, bioaccumulative and toxic substances, vol 2. ACS Books, Washington, DC, pp 184–202
56. Muir D, Braekevelt E, Tomy G, Whittle M (2002) Analysis of medium chain chlorinated paraffins in Great Lakes food webs and in a dated sediment core from Lake St. Francis in the St. Lawrence River system. Preliminary report to Existing Substances Branch, Environment Canada, Hull, Quebec. 9 pp
57. Cooley HM, Fisk AT, Weins SC, Tomy GT, Evans RE, Muir DCG (2001) Examination of the behavior and liver and thyroid histology of juvenile rainbow trout (*Oncorhynchus mykiss*) exposed to high dietary concentrations of C_{10}, C_{11}, C_{12} and C_{14} polychlorinated alkanes. Aquat Toxicol 54:81–99
58. Marvin CH, Painter S, Tomy GT, Stern GA, Braekvelt E, Muir DCG (2003) Spatial and temporal trends in short-chain chlorinated paraffins in Lake Ontario sediments. Environ Sci Technol 37:4561–4568
59. Bezchlebová J, Černohláková J, Kobetičová K, Lána J, Sochová I, Hofman J (2007) Effects of short-chain chlorinated paraffins on soil organisms. Ecotoxicol Environ Saf 67:206–211

60. Sverdrup LE, Hartnik T, Mariussen E, Jensen J (2006) Toxicity of three halogenated flame retardants to nitrifying bacteria, red clover (*Trifolium pratense*) and a soil invertebrate (*Enchytraeus crypticus*). Chemosphere 64:96–103
61. CSTEE (1998) Scientific committee on toxicity, ecotoxicity and the environment opinion on the risk assessment of short chain length chlorinated paraffins. Available at http://ec.europa.eu/health/ph_risk/committees/sct/docshtml/sct_out23_en.htm
62. Sochová I, Hofman J, Holoubek I (2007) Effects of seven organic pollutants on soil nematode *Caenorhabditis elegans*. Environ Int 33:798–804
63. Report from 4th meeting of the POPs Review Committee UNEP/POPS/POPRC.4/10
64. UNEP (2009) Bioconcentration study of a chlorinated paraffin (C=13). Submitted by the Government of Japan. UNEP/POPS/POPRC.5/INF/23

Synthesis of Polychloroalkanes

Vladimir A. Nikiforov

Abstract This chapter describes the preparation of polychloroalkanes (chlorinated paraffines, PCA) as analytical standards. A list of commercially available PCA standards is given. General methods for the synthesis of PCAs are reviewed. Possibilities of the synthesis of other PCA congeners as standards are discussed. References to the original syntheses are tabulated. A selection of synthetic procedures is collected in the Appendix.

Keywords Chlorinated paraffines, Chlorine, Polychloroalkanes, Synthesis

Contents

V.A. Nikiforov
Department of Chemistry, St.Petersburg State University, Universitetskiipr 26, 198504 St.Petersburg, Russia
e-mail: vovanikiforov@yahoo.co.uk

J. de Boer (ed.), *Chlorinated Paraffins*, Hdb Env Chem (2010) 10: 41–82,
DOI 10.1007/698_2009_40, © Springer-Verlag Berlin Heidelberg 2010,
Published online: 16 February 2010

1   Introduction

The first goal of this review is to describe to the community of environmental analytical chemists how the polychlorinated alkane (PCA) or chlorinated paraffin (CP) compounds they use for calibration and other purposes are prepared. This was an achievement of the last decade and therefore, it will be a chronological description with personal tribute to those few good men and women who have made it possible.

The second goal is to review the general methods for synthesis of PCA developed in the past century, to look into other synthetic possibilities, to provide some generalizations and some new ideas for those, who might undertake the synthesis of further PCA congeners. This part will be organized on the chemistry basis.

There will be little or no synthetic details on the figures, except where they are really important. There will be very few indications of preparative yields; however, it is worth emphasizing that all the reactions mentioned can be useful in practice. A selection of preparative methods forms an appendix. Nevertheless, a chemist planning synthesis of PCAs is advised to study original papers and the references therein carefully.

2   Synthesis of Polychloroalkanes as Standards
for Environmental Analytical Chemistry

The first synthesis of a PCA as analytical standard – 1,2,5,6,9,10-hexachlorodecane – was reported in 1997 by Tomy and co-workers [1, 2]. Authors employed addition of chlorine to commercially available 1,5,9-decatriene (Fig. 1).

The sample contained a number of impurities; however, it was used for quantification purposes in the analysis of C_{10}–C_{13} PCA technical mixtures.

In 1998, Fisk and co-workers [3, 4] reported synthesis of a series of tetra- to octachloro C_{10}, C_{11}, and C_{14} alkanes with full or partial structure elucidation. PCA congeners were obtained by chlorination of terminal alkadienes and 1,5,9-decatriene (Fig. 2). None of the compounds were isolated in pure form.

Electrophilic addition of chlorine to double bonds was accompanied by free-radical substitution. This led to over-chlorinated by-products. However, several of

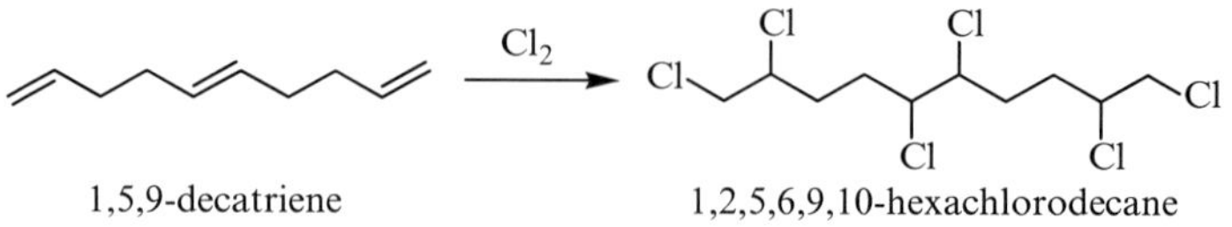

Fig. 1 First synthesis of a polychloroalkane for the needs of environmental chemistry by Tomy and co-workers [2]

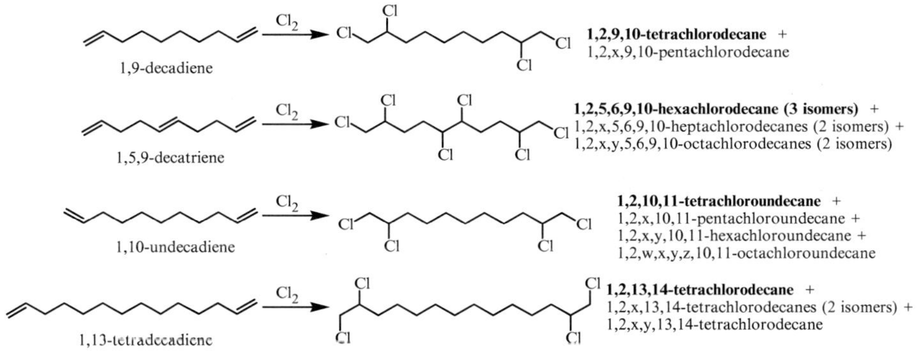

Fig. 2 Synthesis of PCA congeners by chlorination of alkadienes and 1,5,9-decatriene [4]

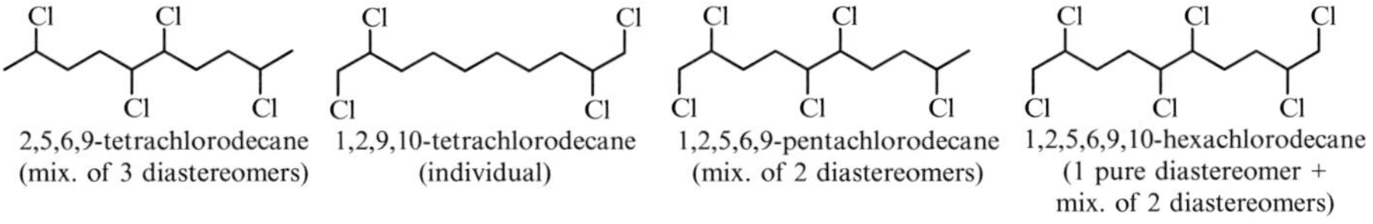

Fig. 3 Polychlorodecanes synthesized by Coelhan and co-workers [5]

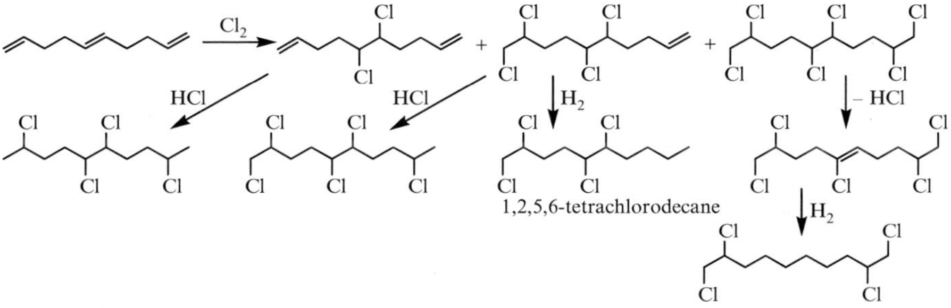

Fig. 4 Synthesis of polychlorodecanes from 1,5,9-decatriene by use of chlorination, hydrochlorination, dehydrochlorination and hydrogenation reactions [6]

those were resolved with GC, identified by their mass-spectra and multi-component mixtures were used for quantification.

In 1998, Coelhan and co-workers [5] reported a series of tetra-, penta-, and hexachlorodecanes (Fig. 3) without giving synthetic details. It is easy to assume that the author used the same commercially available 1,5,9-decatriene, and compounds were obtained by the combination of chlorination and hydrochlorination of double bonds.

In the following work (2003), Coelhan reported synthesis of the aforementioned and other PCAs [6] (Figs. 4 and 5).

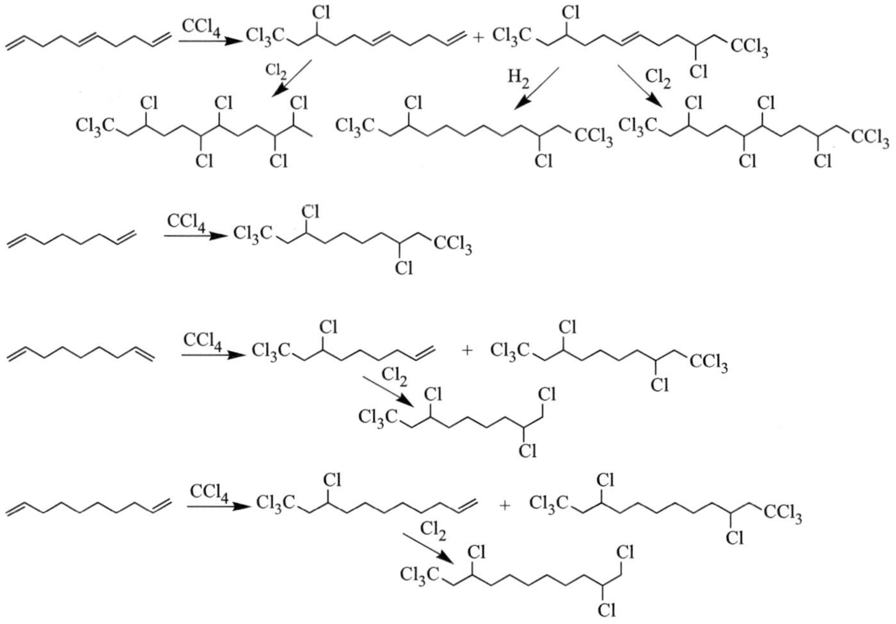

Fig. 5 Synthesis of C_{10}–C_{12} polychloroalkanes (Cl_6–Cl_{10}) via addition of tetrachloromethane, chlorine or hydrogen to double bonds [6]

In this work, a more diverse chemistry was applied. This resulted in a larger variety of PCA structures. Not only the addition of chlorine and hydrogen chloride but also the free-radical addition of CCl_4 was further explored. It is well-known that CCl_4 adds to terminal double bonds selectively: CCl_3 always adds to a terminal carbon atom. Like in previous works, many isomers were obtained as mixtures of diastereomers. Some of them were separated and isolated in individual state.

In PCA's, each CHCl group is a center of chirality; therefore the number of diastereomers can be as high as 2^{n-1}, where n is a number of chiral centers. Stereoselectivity aspects will be discussed in Sect. 3.1.

Moreover, the starting compound, commercial 1,5,9-decatriene, is a mixture of *cis*- and *trans*-isomers, and this theoretically may double the number of diastereomers formed.

In 2000, Tomy and coworkers [7] reported chlorination of individual PCAs C_{10}–C_{14}. Mixtures of congeners, containing 4–9 Cl atoms per molecule were used for quantification purposes in the analysis of environmental samples.

In 2005 and 2006, Beaume [8, 9] reported synthesis of 5 other individual decachlorobornanes. 1,2,4,5,9,10-hexachlorodecane was obtained by the addition of three molecules of Cl_2 to three double bonds of 1,4,9-decatriene (Fig. 6).

In this synthesis, the author used the pure *trans*-isomer of the starting compound and as a result, the final product was a mixture of just two diastereomers.

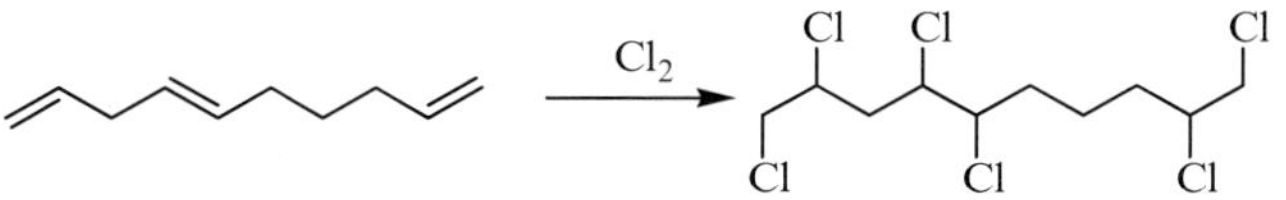

1,4-*trans*-9-decatriene   1,2,4,5,9,10-hexachlorodecane
                                        2 diastereomers

Fig. 6 Addition of chlorine to 1,4-*trans*,9-decatriene leading to isolation of a mixture of 2 diastereomers of 1,2,4,5,9,10-hexachlorodecane [8]

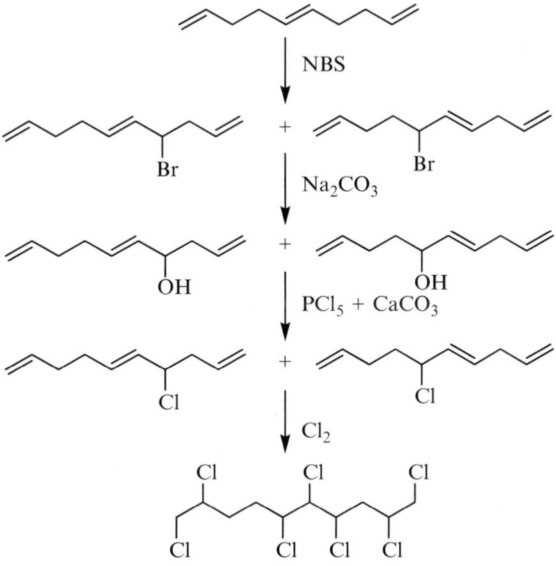

Fig. 7 A sequence of free-radical bromination, hydrolysis, reaction with PCl₅ and addition of Cl₂ to double bonds yields 1,2,4,5,6,910-heptachlorodecane [8]

1,2,4,5,6,9,10-heptachlorodecane

Selective allylic bromination of 1,4-*trans*,9-decatriene with N-bromosuccinimide (NBS) led to a mixture of two isomeric monobromides. The mixture was hydrolyzed to the mixture of corresponding alcohols and then converted to the mixture of two monochlorodecatrienes. The addition of 3 eq. of chlorine yielded 1,2,4,5,6,9,10-heptachlorodecane (Fig. 7).

Formation of isomers in free-radical allylic substitution is a general rule. In this case, abstraction of a hydrogen atom from C4 of the parent molecule leads to the formation of a delocalized allylic radical, with spin density distributed between two carbon atoms C-4 and C-6. Then this radical abstracts the bromine atom from NBS and adds it to one or the other position (Fig. 8):

The inevitable formation of isomers should be taken into account when planning syntheses of PCA involving free-radical allylic substitution.

Another isomer of heptachlorodecane was synthesized from the same starting material [8]. The addition of 1 eq. of chlorine proceeded with good selectivity and gave 5,6-dichlorodecadiene-1,9. It is a general rule that internal double bonds are

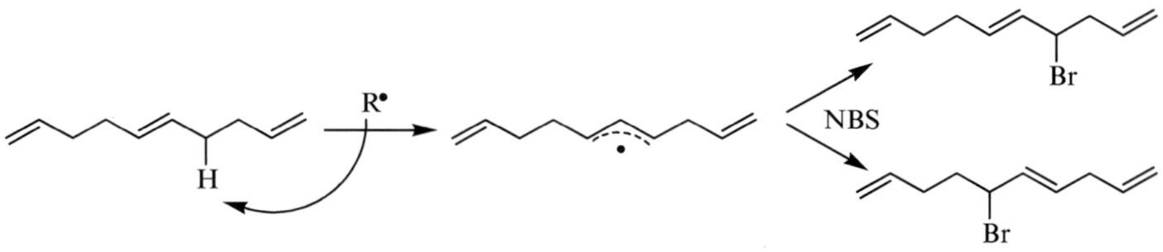

Fig. 8 Formation of intermediate allylic radical leads to a mixture of isomeric bromides

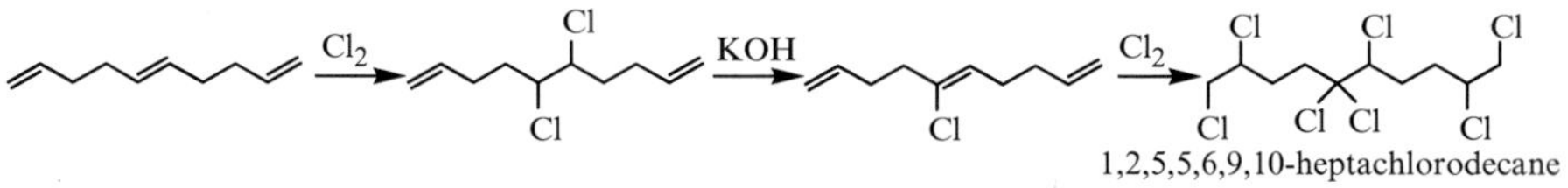

Fig. 9 Synthesis of 1,2,5,5,6,9,10-heptachlorodecane: application of chlorination-dehydrochlori-
nation-chlorination sequence allows conversion of C=C internal double bond into –CCl$_2$-CHCl-
group [8]

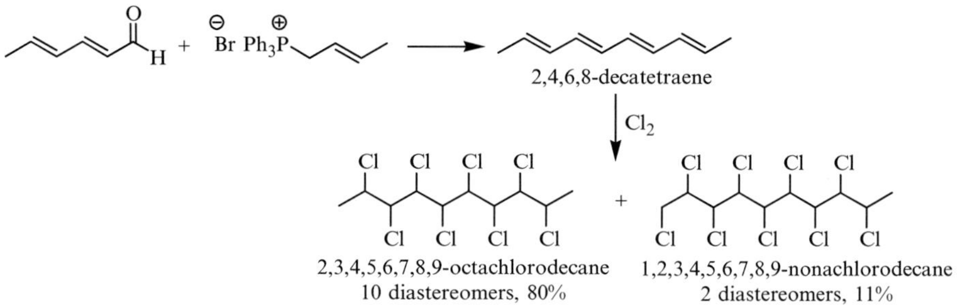

2,4,6,8-decatetraene

2,3,4,5,6,7,8,9-octachlorodecane   1,2,3,4,5,6,7,8,9-nonachlorodecane
10 diastereomers, 80%   2 diastereomers, 11%

Fig. 10 Construction of a non-chlorinated C$_{10}$ precursor from two smaller molecules and synthesis
of octa- and decachlorodecanes [8]

more reactive to electrophilic agents (Cl$_2$, HCl) than terminal double bonds.
Dehydrochlorination of 5,6-dichlorodecadiene-1,9 yielded 5-chlorodecatriene-
1,5,9 and this compound was converted to 1,2,5,5,6,9,10-heptachlorooctane by
the addition of 3 eq. of chlorine (Fig. 9).

For the synthesis of octa- and nonachlorodecanes, the authors prepared 2,4,6,8-
decatetraene [8, 9]. Its chlorination followed by separation resulted in an 80% yield
of 2,3,4,5,6,7,8,9-octachlorodecane (a mixture of ten (!) diastereomers) and two
diastereomers of 1,2,3,4,5,6,7,8,9-nonachlorodecane in 8% and 3% yield (Fig. 10).

As described in the aforementioned report [4], the addition of chlorine to double
bonds was accompanied by allylic substitution.

Crystal structure of one of the compounds, a single diastereomer of 1,2,5,6,9,10-
hexachlorodecane was reported by Frenzen [10] (Fig. 11).

Two other reports on the direct structural resolution of PCA appeared in 1974 and
in 1979 [11, 12]: (*RRSS*)-2,3,6,7-tetrachlorooctane and *meso-(RS)*-1,1,1,3,6,8,8,8-
Octachlorooctane (Fig. 11).

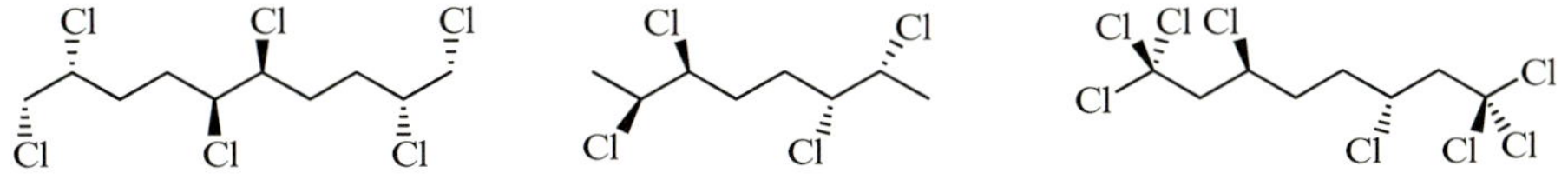

Fig. 11 Configuration of polychloroalkanes from X-ray diffraction data. From left to right: (2SR,5RR,6RR,9SR)-1,2,5,6,9,10-hexachlorodecane [10], (*RRSS*)-2,3,6,7-tetrachlorooctane [12] and *meso-(RS)*-1,1,1,3,6,8,8,8-Octachlorooctane [11]

These are important reference points for the structure elucidation of individual PCAs in the future.

A number of PCA congeners are commercially available through a variety of suppliers. However, it appears that they all originate from just two sources: Chiron, Norway and Dr. Ehrenstorfer GmbH, Germany. Compounds available from Dr. Ehrenstorfer are those reported in [5, 6, 8, 9], and compounds from Chiron are their own production. Synthetic methods used by Chiron are not disclosed, but one can easily see that they are produced by the addition of CCl_4 or Cl_2 to terminal n-alkenes or n-alkadienes. The 48 individual PCA standards are listed in the Table 1.

The number of C atoms varies between 8 and 20, and the number of Cl atoms between 1 and 9.

A large number of monochloroalkanes, several dichloroalkanes, and some polychloroalkanes with short chain lengths are available from suppliers of general chemicals.

Three major conclusions can be made regarding the synthesis of PCA congeners as analytical standards:

1. Formation of diastereomers is a major obstacle to pure PCA congeners. Each isomer can be a mixture of up to 2^{n-1} diastereomers, where n is the number of chiral centers(–CHCl-groups).
2. Otherwise, with a variety of methods already explored, one can prepare virtually any congener.
3. However, it is not clear at present which individual PCA congeners, if any, need to be synthesized. There are no reports in the literature on any single congener that is predominant in technical formulations or environmental samples, or particularly persistent, bioaccumulative, or toxic.

3   General Methods of Syntheses of Polychloroalkanes

There are a large number of reports on the synthesis of polychloroalkanes. Many of the methods can be useful in the preparation of PCA congeners as standards. Because of limited space, only the methods applied to alkanes with more than 8 carbon atoms will be discussed.

Table 1 Commercially available PCA standards

Number of C	Number of Cl	PCA congener name	Source[a]
8	2	1,2-Dichlorooctane	C
8	4	1,1,1,3-Tetrachlorooctane	C
8	4	1,2,7,8-Tetrachlorooctane	C
8	8	1,1,1,3,6,8,8,8-Octachlorooctane	C
9	2	1,2-Dichlorononane	C
9	4	1,1,1,3-Tetrachlorononane	C
9	4	1,2,8,9-Tetrachlorononane	C
9	6	1,1,1,3,8,9-Hexachlorononane	C
10	2	1,2-Dichlorodecane	C
10	4	1,1,1,3-Tetrachlorododecane	C
10	4	2,5,6,9-Tetrachlorododecane (mix of 3 diastereomers)	C, E
10	4	1,2,9,10-Tetrachlorododecane (1 diastereomer)	E
10	5	1,2,5,6,9-Pentachlorododecane (mix of 2 diastereomers)	E
10	6	1,1,1,3,9,10-Hexachlorodecane (isomer mixture)	C
10	6	1,2,5,6,9,10-Hexachlorododecane (1 diastereomer)	E
10	6	1,2,5,6,9,10-Hexachlorododecane (mix of 2 diastereomers)	E
10	6	1,2,4,5,9,10-Hexachlorododecane	E
10	7	1,2,4,5,6,9,10-Heptachlorododecane	E
10	7	1,2,5,5,6,9,10-Heptachlorododecane	E
10	8	1,1,1,3,8,10,10,10-Octachlorodecane	C
10	8	2,3,4,5,6,7,8,9-Octachlorododecane	E
10	9	1,2,3,4,5,6,7,8,9-Nonachlorododecane	E
11	2	1,2-Dichloroundecane	C
11	4	1,1,1,3-Tetrachloroundecane	C
11	4	1,2,10,10-Tetrachloroundecane	C
11	6	1,1,1,3,10,11-Hexachloroundecane (isomer mixture)	C
11	8	1,1,1,3,9,11,11,11-Octachloroundecane	C
12	2	1,2-Dichlorododecane	C
12	2	1,12-Dichlorododecane	C
12	4	1,1,1,3-Tetrachlorododecane	C
12	4	1,2,11,12-Tetrachlorododecane	C
12	6	1,1,1,3,11,12-Hexachlorododecane (isomer mixture)	C
12	8	1,1,1,3,10,12,12,12-Octachlorododecane	C
13	2	1,2-Trichlorotridecane	C
13	4	1,1,1,3-Trichlorotridecane	C
13	6	1,1,1,3,12,13-Hexachlorotridecane (isomer mixture)	C
13	8	1,1,1,3,11,13,13,13-Octachlorotridecane	C
14	2	1,2-Dichlorotetradecane	C
14	4	1,1,1,3-Tetrachlorotetradecane	C
14	4	1,2,13,14-Tetrachlorotetradecane	C
14	8	1,1,1,3,12,14,1,4,14-Octachlorotetradecane	C
15	6	1,1,1,3,14,15-Hexachloropentadecane	C
16	8	1,1,1,3,14,16,16,16-Octachlorohexadecane	C
17	8	1,1,1,3,15,17,17,17-Octachloroheptadecane	C
18	1	1-Chlorooctadecane	C
18	8	1,1,1,3,16,18,18,18-Octachlorooctadecane	C
19	8	1,1,1,3,17,19,19,19-Octachlorononadecane	C
20	8	1,1,1,3,18,20,20,20-Octachloroeicosane	C

[a]C Chiron; E Dr. Ehrenstorfer

3.1 General Methods for Introduction of Cl Atoms in Place of Functional Groups

3.1.1 Addition of HCl to Double Bonds

Electrophylic addition of HCl to double bonds was one of the first methods for the synthesis of chloroalkanes. Addition to terminal double bonds occurs regio-selectively with the formation of 2-chloroalkane as a mixture of two enantiomers (Fig. 12). Addition to internal double bonds gives mixtures of isomers in the case of asymmetrical alkene (both as racemic mixtures of enantiomers, Fig. 12):

3.1.2 Addition of Cl₂ to Double Bonds

Addition of chlorine to terminal double bonds, like addition of HCl, gives single compounds as a racemic mixture of enantiomers. Chlorine is usually more reactive than hydrogen chloride. Internal double bonds are usually more reactive towards chlorine than terminal double bonds. The addition of chlorine to internal double bonds gives mixtures of four diastereomers (*RR, RS, SR* and *SS*, Fig. 13).

Under certain conditions, addition occurs as mainly *syn-* or mainly *anti-*. Therefore, the use of pure *cis-* or pure *trans-*isomer of an alkene will limit the number of products as shown for *anti-*addition of chlorine to *trans-*olefin (Fig. 14).

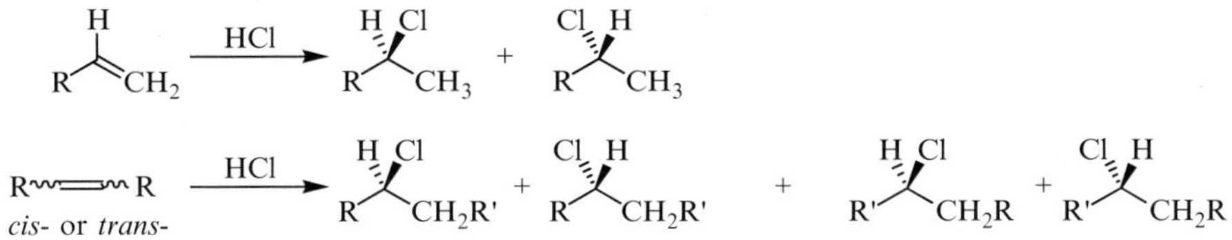

Fig. 12 Formation of enantiomers by the addition reaction of HCl to C=C double bonds

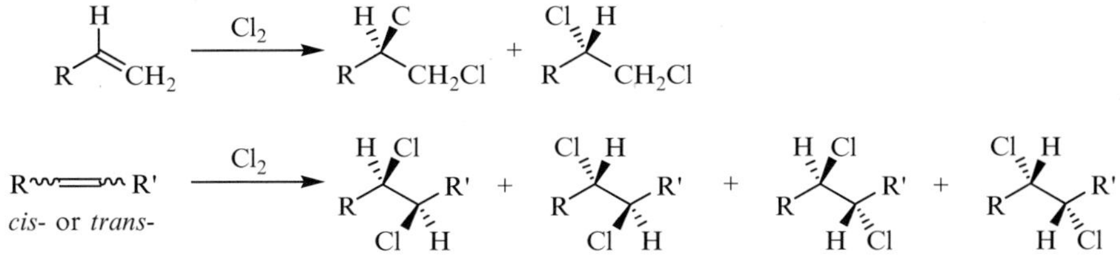

Fig. 13 Non-stereospecific addition of Cl₂ to terminal and internal double bonds

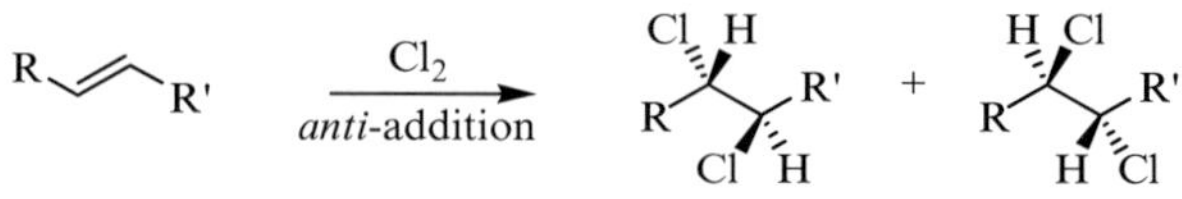

Fig. 14 Possibility of stereospecific (*anti*) addition of Cl_2 to double bonds

Fig. 15 Stereoselective conversion of alcohols to chlorides with the same configuration (by reaction with $SOCl_2$ in the absence of base) and with the opposite configuration (by reaction with $SOCl_2$ in presence of base)

Fig. 16 Ring-opening of the cyclic ether with formation of 1,5-dichlorooctane by action of $SOCl_2$ [13]

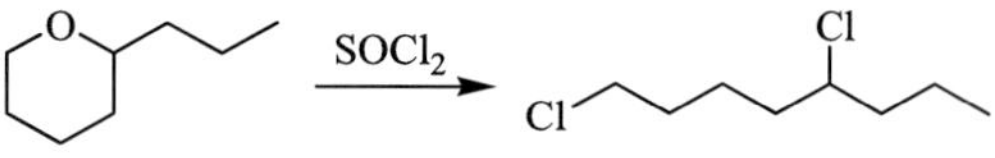

3.1.3 Substitution of OH to Cl

There are a number of ways to substitute hydroxylic groups with Cl. The simplest is the reaction with HCl in the presence of an acid catalyst, for example $ZnCl_2$. Reaction with tionyl chloride, $SOCl_2$ is more convenient, as it often allows stereo-selective transformation and provides the opportunity to obtain both enantiomers of the product with either the same or the opposite configuration (Fig. 15):

Alkoxy-groups, for instance –OMe, –OEt can also be substituted for Cl. This reaction is applicable to cyclic ethers; it was used for the preparation of 1,5-dichlorooctane [13] (Fig. 16).

3.1.4 Conversion of C=O to CCl_2

Phosphorous pentachloride is the traditional reagent for the transformation of C=O to CCl_2. However, the reaction is accompanied by the formation of mixtures of chloro-olefins; often chloroolefins are the main products. Compounds with a –CH=CCl– fragment can then be converted to polychloroalkanes (Fig. 17).

Alternatively, carbonyl compounds can be converted to *gem*-dichlorides through formation of oximes followed by chlorination (Fig. 17). This method has already been used for the synthesis of 2,2-dichlorononane [14].

An interesting approach can be suggested for the future syntheses of single diastereomers/enantiomers of PCA: the use of natural compounds with high diaste-reomeric/enantiomeric purity as precursors, as shown below for a hypothetical natural compound (Fig. 18).

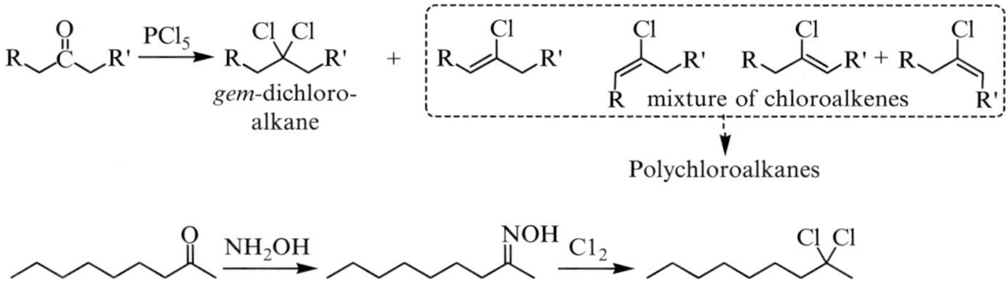

Fig. 17 Preparation of *gem*-dichloroalcanes from carbonyl compounds directly by reaction with PCl$_5$ or more selective preparation via conversion to oximes with subsequent chlorination [14]

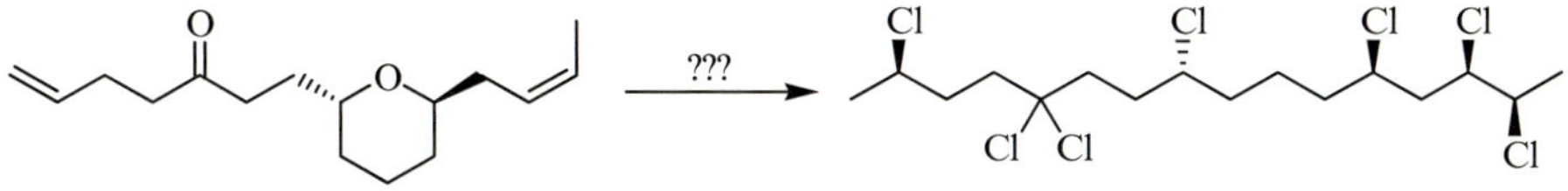

Fig. 18 Suggestion for future synthesis: Stereospecific conversion of a hypothetical natural product to a single enantiomer of a 2,5,5,8,12,14,15-heptachlorohexadecane

3.2   Substitution of H to Cl: Free-Radical Chlorination of Alkanes or Lower Chlorinated Alkanes

Direct chlorination of alkanes leads to complex mixtures of isomers. Regioselectivity is low and different H-atoms with comparable rates are substituted. Moreover, introduction of a Cl atom has no major effect on the reactivity. Therefore, further chlorination of monochloroalkanes to dichloroderivatives will occur to a large extent. This general rule was confirmed by Tomy and coworkers [2, 4, 7]; chlorination of n-alkanes yielded mixtures of congeners with varying degrees of chlorination. No single peaks were resolved in a gas chromatogram.

Careful monochlorination of 1-chlorooctane produced mixtures of dichloroderivatives, which were successfully resolved by GC [13]. 1,X-dichlorooctanes eluted in the order of increase of X. Authentic samples of 1,1-, 1,2-, 1,4-, and 1,5-dichlorooctanes were synthesized for identification purposes (Fig. 19).

A similar approach was applied to the chlorination of 1,1-di- and 1,1,1-trichlorooctanes [15]. The percentages of 1,1,X-trichlorooctanes and 1,1,1,X-tetrachlorooctanes formed are shown in Fig. 20.

1,1,8-Trichlorooctane and a mixture of 1,1,2- and 1,1,3-trichlorooctanes were isolated by preparative GC. Pure isomers and isomer mixtures were analyzed by ^{1}H NMR.

Recently, a similar procedure was applied to 1-chlorodecane [16]. Based on an experimentally determined composition of initial chlorination products, the authors developed a Monte-Carlo model for prediction of the composition of PCA mixtures. There was also an early attempt to build a model for prediction of PCA

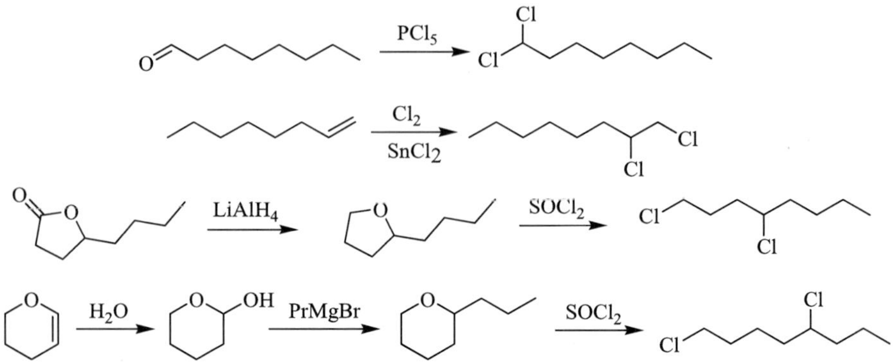

Fig. 19 Synthesis of authentic samples of 1,1-, 1,2-, 1,4- and 1,5-dichlorooctanes [13]

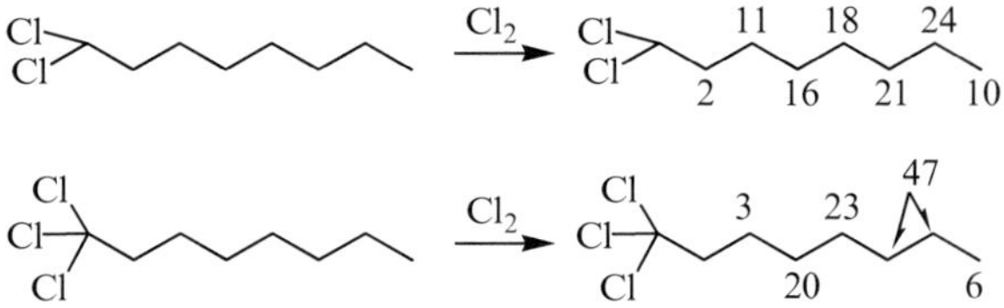

Fig. 20 Percentages of monochlorination products, obtained by direct chlorination of 1,1-di- and 1,1,1-trichlorooctanes [15]

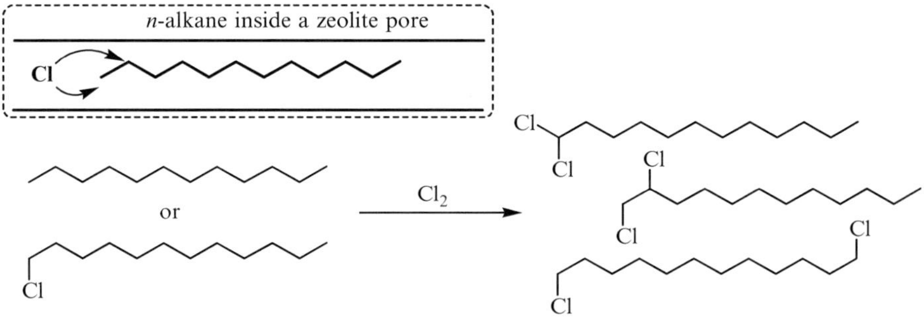

Fig. 21 Selective chlorination of an encapsulated precursor at near-end positions of carbon chain [18, 19]

mixture compositions based on experimentally determined reaction rates for mono- and dichlorination of n-dodecane [17].

Generally speaking, free-radical chlorination is not a method for the preparation of single PCA congeners. However, there is an interesting approach that allows selective chlorination: encapsulation of substrates on zeolites [18, 19]. It has been shown that chlorination of a dodecane molecule absorbed on zeolites occurs only near the end of the alkyl chain. Obviously, pore size does not allow the chlorine molecules (or Cl• radical) to reach the central part of a long carbon chain (Fig. 21).

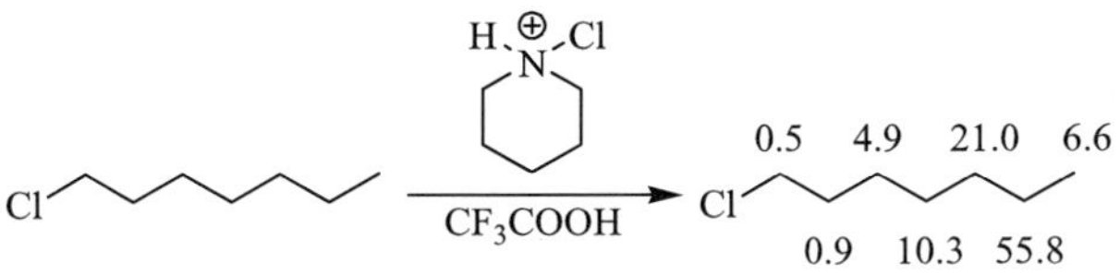

Fig. 22 Selectivity of chlorination of 1-chloroheptae with N-chloropiperidine in trifluoroacteic acid [21]. Numbers show percentages

Monochlorination of dodecane on zeolites produces mainly 1- and 2-chlorododecanes. Further chlorination yields mixtures of dichlorododecanes. The composition of the dichlorodcrivativcs depends on the zeolite used and varies as follows: 1,1-dichlorododecane (28–40%), 1,2-dichlorododecane (33–43%), 1,12-dichlorododecane (12–19%). Monochlorination of 1-chlorododecane gives 1,1-dichlorododecane (13%), 1,2-dichlorododecane (5%), 1,12-dichlorododecane (49%) (Fig. 21).

It is not known whether polychloroalkanes can be absorbed on zeolites; if that is the case, chlorination on zeolites can be used for selective introduction of one or two additional Cl atoms into terminal groups of the carbon chain.

Another selective chlorination agent is N-chloropiperidine. [20, 21]. The authors studied the chlorination of 1-X-alkanes (C_3–C_7) with N-chloropiperidine in trifluoroacetic acid. It was seen that the reagent is very sensitive to the electron-withdrawing effect of substituent X. In the case of chlorination of 1-chloropropane, the following relative content of dichlorides was found (Fig. 22):

Chlorination of the most distant CH_2 group accounts for more than 50% of the product mixture. This is explained by the highly electrophilic nature of the reacting radical – piperidine radical-cation. This method can be recommended for selective introduction of one Cl atom into a molecule of a polychloroalkane. The new Cl atom will enter the CH_2 group, the most distant from the other Cl atoms already present in the molecule.

3.3 Kharasch Addition

The addition of tetrachloromethane and similar compounds was discovered by Kharasch in 1945 [22–26]. The main difference between this reaction and the general modification of the carbon chain is that the number of C atoms in the product increases. Initially, tetrachloromethane and chloroform were used. It was found that CCl_3 adds to C-1 of 1-alkene, while Cl or H adds to C-2. The initiator for the reaction was benzoyl peroxide. Therefore, it was concluded that this reaction is a chain free-radical process (Fig. 23).

Kharasch also reported bis-addition of CCl_4 to 1,5-hexadiene and synthesis of 1,1,1,3,6,8,8,8-octachlorooctane with good yield (Fig. 24). Therefore, the reaction can be used for the formation of polychloroalkane with a carbon chain that is two atoms longer than in the alkadiene precursor.

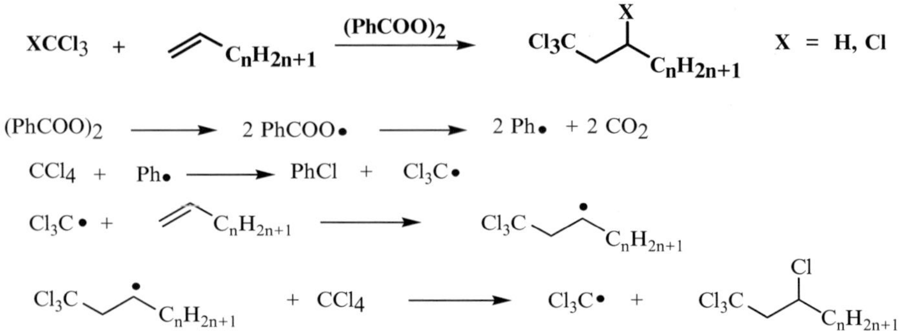

Fig. 23 Kharasch addition: reaction of tetrachloromethane or chloroform with 1-alkenes and free-radical mechanism of the chain reaction with CCl_4 [22, 26]

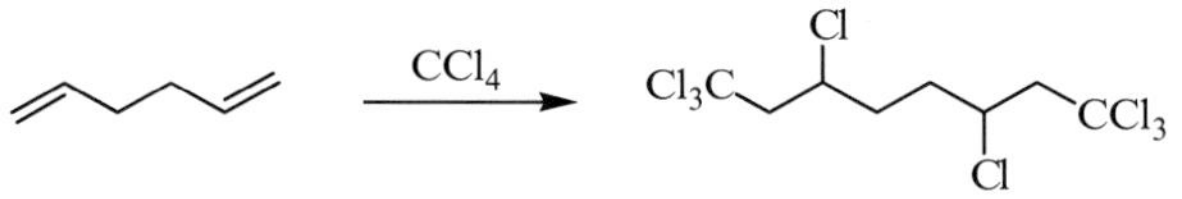

Fig. 24 Kharasch bis-addition of CCl_4 to 1,5-hexadiene: synthesis of 1,1,1,3,6,8,8,8-Octachlorooctane [22]

Later, it was found that other compounds with a trichloromethyl group react with terminal carbon-carbon double bonds [27–32]. The studied reagents included 1,1,1-trichloroethane, 1,1,1,2,2-pentachloroethane, 1,1,1-trichloropropane, 1,1,1,3,3-pentachloropropane, and 1,1,1,3-tetrachloropropane - the latter itself being a product of the Kharasch addition of carbon tetrachloride to ethylene (Fig. 25).

Kharasch addition to 1-alkenes is a general synthetic pathway to 1,1,1-trichloroalkanes, 1,1,1,3-tetrachloroalkanes, 2,2,4-trichloroalkanes, 3,3,5-trichloroalkanes, and 1,3,3,5-tetrachloroalkanes. Bis-addition to C_N 1,(N-1)-alkadienes is a way to 1,1,1,N,N,N-hexachloroalkanes, 1,1,1,3,(N-2),N,N,N-octachloroalkanes, 2,2,4,(N-3),(N-1),(N-1)-hexachloroalkanes, 3,3,5,(N-4),(N-2),(N-2)-hexachloroalkanes, and 1,3,3,5,(N-4),(N-2),(N-2),N-octachloroakanes. Addition of 1,1,1,5-tetrachloropentane to ethylene and propylene has also been reported [33]. This is an example of the use of the CCl_3 component with the longest carbon chain (Fig. 26).

Unfortunately, CCl_4 has been the only CCl_3 component of the Kharasch synthesis used so far for the preparation of PCA congeners for the needs of environmental chemistry [6].

On the other hand, double-bonds containing components of the Kharasch synthesis are not necessarily simple alkenes or alkadienes only, but could also be already chlorinated compounds [29, 33–35]. Examples of such compounds include vinylchloride, vinylidenedichloride, allylchloride, and 2-chloropropene (Fig. 27).

The addition of 1,1,1-trichloroethane and 1,1,1,3-tetrachloropropane to conjugated diene, 1,3-butadiene results in 1,4-conjugated addition and formation of an unsaturated polychlorocompound [36] (Fig. 28).

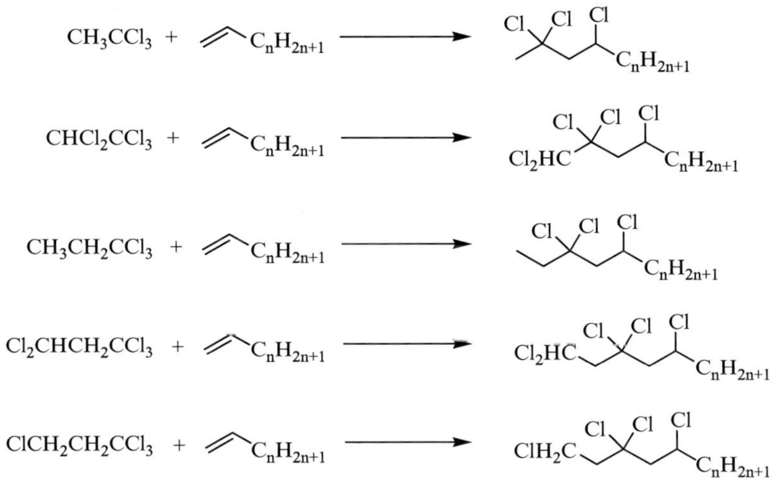

Fig. 25 Kharasch addition of 1,1,1-trichloroethane, 1,1,1,2,2-pentachloroethane, 1,1,1-trichloropropane, 1,1,1,3,3-pentachloropropane and 1,1,1,3-tetrachloropropane to alkenes [27–32]

$$ClCH_2CH_2CH_2CH_2CCl_3 + CH_2{=}CHR \longrightarrow ClCH_2CH_2CH_2CH_2CCl_2CH_2CHClR \quad (R = H, CH3)$$

Fig. 26 Adition of 1,1,1,5-tetrachloropentane to ethylene and propylene [33]

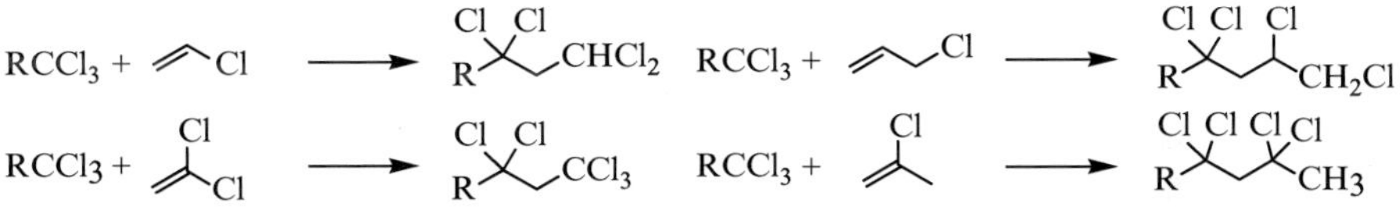

Fig. 27 Kharasch addition to chlorinated alkenes: vinylchloride, vinylidenedichloride, allylchloride and 2-chloropropene [29, 33–35]

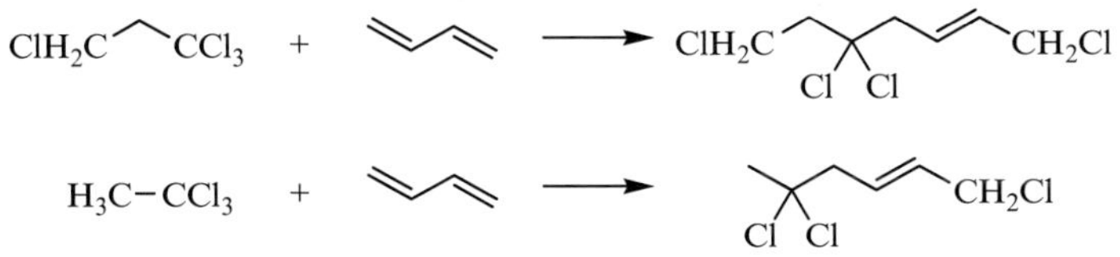

Fig. 28 Addition of 1,1,1-trichloroethane and 1,1,1,3-tetrachloropropane to 1,3-butadiene [36]

Polychloroalkenes obtained in this way can be converted to polychloroalkanes with two more, one more or the same number of chlorine atoms via chlorination, hydrochlorination, or hydrogenation of the double bond, respectively. It was shown by Coelhan [6] that the presence of distant double bonds in alkadienes and

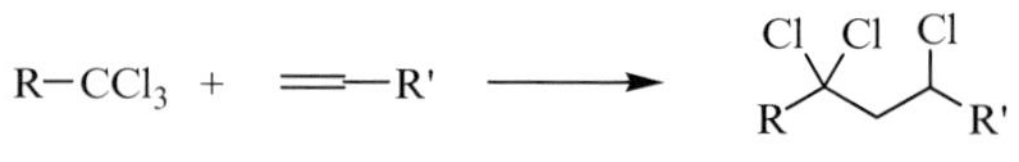

Fig. 29 General application of the Kharasch addition as a way to synthesize polychloroalkanes (R, R' = linear radical with/without double bonds and/or Cl substituents)

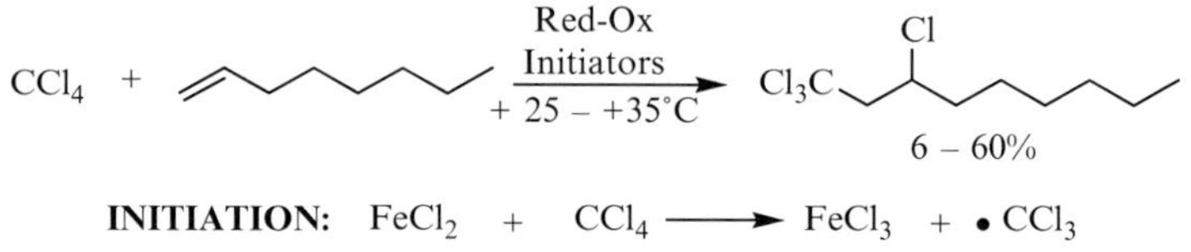

Fig. 30 Use of Red-Ox initiators in the Kharasch reaction; mechanism of initiation in case of $FeCl_2$ [39]

alkatrienes does not affect the Kharasch addition. Moreover, it is possible to isolate mono- as well as bis-adducts. The variety of structurally different $-CCl_3$ and $CH_2=CH-$ components from the Kharasch addition reactions permits conclusions to be drawn and gives a practical advice: any compound with a CCl_3 group and other Cl substituents and/or double bonds may give practical yields of Kharasch adducts with any 1-olefin, containing other double bonds and/or Cl substituents (Fig. 29).

This opens the way to a variety of PCA congeners for environmental chemistry needs via an unexplored approach: construction of a PCA molecule from two parts of similar size; in other words, construction of a long-chain polychloroalkane or polychloroalkane precursor from two shorter-chain polychloroalkane and poly-chloroalkene derivatives.

There are numerous publications on the improvements to the Kharasch reaction. The primary method for improvement was the introduction of more sophisticated initiators (instead of peroxides [22–26, 37]). It was shown that various Red–Ox type initiators, like metal salts in different oxidation states (Cu^+, Cu^{2+}, Fe^{3+}, etc.), salts with organic acids (Co acetoacetate, Mn stearate), and a metal carbonyl – $Fe(CO)_5$ give adducts with practical yields [38, 39] and allow to carry the reaction at "near room" temperatures (Fig. 30).

For example, abstraction of a Cl atom from CCl_4 by $FeCl_2$ gives $FeCl_3$ and a CCl_3 radical, thus initiating a chain reaction as shown in Fig. 23. In addition to $Fe(CO)_5$ other metal carbonyls or mixed carbonyls with other ligands were used [27, 28, 31, 40–42], including complexes with cyclopentadienyl ligands, analogous to ferrocene [40, 41, 43]. For binuclear complexes of Fe, Mo, and Ru a non-radical mechanism was suggested [40, 43] that includes a catalytic cycle with a series of oxidative addition/reductive elimination steps in a metal coordination sphere (Fig. 31).

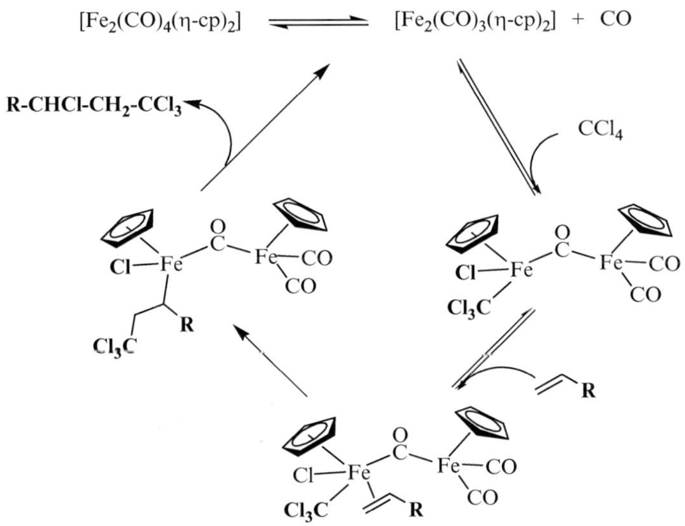

Fig. 31 Suggested non-radical mechanism of addition of carbon tetrachloride to 1-alkenes [43]

$$n\text{-C}_6\text{H}_{13}\text{-CH=CH}_2 + \text{CCl}_4 \xrightarrow{[\text{Ru}]} n\text{-C}_6\text{H}_{13}\text{-CHCl-CH}_2\text{CCl}_3 \quad \textbf{Yield = 85\%}$$

$$\textbf{CE = 4250}$$

$$[\text{Ru}] = \text{RuH}_3(\text{SiR}_3)(\text{PPh}_3)_3, \quad \text{RuH}(\text{SiR}_3)(\text{PPh}_3)_2$$

Fig. 32 High efficiency of ruthenium catalysts in the synthesis of polychloroalkanes [45]. CE is catalytic efficiency

However, parallel reactions by two mechanisms (metal-mediated and chain free-radical) cannot be excluded [43]. Very efficient catalysts are ruthenium complexes with Si-containing ligands [44, 45] (Fig. 32).

With 0.02 mol% of the most efficient catalyst RuH$_3$(SiMe$_2$Ph)(PPh$_3$)$_3$ the yield of 1,1,1,3-tetrachlorononane is 85% and the catalytic efficiency (CE) (moles of obtained adduct)/(moles of applied catalyst) reaches 4,250. This means that just a few micrograms of the ruthenium complex are sufficient for the preparation of ca. 50 mg sample of a polychloroalkane for environmental analytical purposes.

In recent years ruthenium complexes, like RuCl$_2$(PPh$_3$)$_3$, RuH$_{1-3}$SiR$_3$(PPh$_3$)$_{2-4}$, or RuH$_2$(PPh$_3$)$_4$ were the most studied catalysts [46–48] both in preparative and mechanistic aspects. Reaction of octene-1 with CCl$_4$ is a typical model reaction. Obviously, interest in addition of CCl$_4$ to long-chain olefins is due to the possibility of preparation of valuable fatty acids via hydrolysis of adducts [49, 50] (Fig. 33).

Individual polychloroalkanes were used as model compounds in studies of thermal decomposition of polyvinylchloride [51].

Metal-based catalytic systems can be immobilized on silicagel [52, 53]. Complexes of Cu, Mn, Co, Ni, Fe, and V catalyze the addition of CCl$_4$ to octene-1 [38, 53]. The only product in all cases is 1,1,1,3-tetrachlorononane, except in the

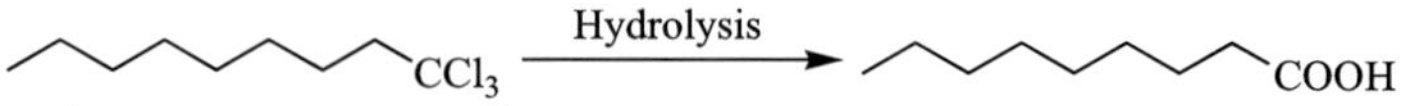

Fig. 33 One of the practical uses of polychloroalkanes with terminal CCl_3-group: conversion to long-chain carboxylic acids [49, 50]

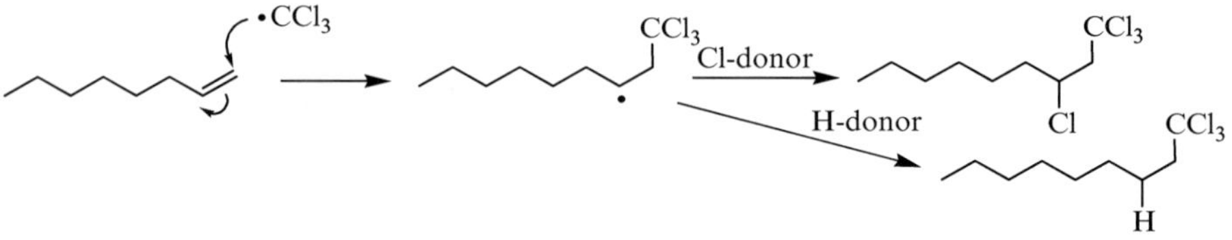

Fig. 34 Formation of both 1,1,1,3-tetrachlorononane and 1,1,1-trichlorononane in reaction of CCl_4 with octene-1 [27, 28, 30]

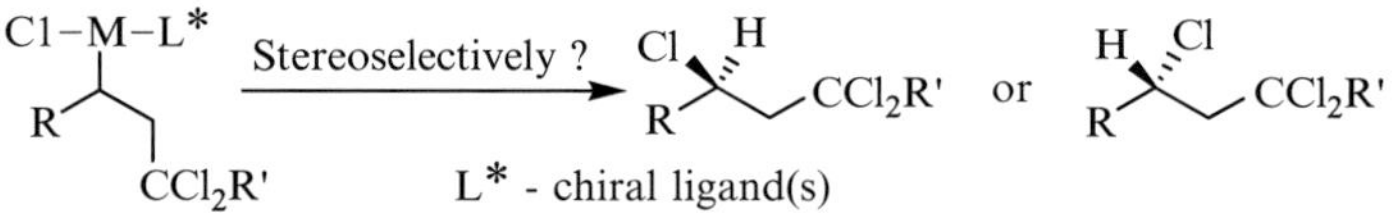

Fig. 35 Possible stereoselective synthesis of polychloroalkanes on metallocomplex catalysts

reaction catalyzed by Fe. In this case telomerization and hydrodechlorination products are formed in noticeable amounts. Partial hydrodechlorination was also observed in homogeneous catalytic reactions on Fe and Mo [27, 28, 30]. The primary by-product is 1,1,1-trichlorononane. Its formation mechanism involves hydrogen abstraction (from solvent or CH_2 groups of alkene) by the intermediate radical (Fig. 34).

This side-reaction is more typical for Fe, than for Mo complexes. Hydrodechlorination can be largely avoided by the addition of good donors of Cl-atoms, like 4,N,N-trichlorobenzenesulfonamide (4-Cl–$C_6H_4SO_2NCl_2$) [28]. Non-metal Red–Ox catalysts are also effective. Good yields of 1,1,1,3-tetrachlorononane were demonstrated for aromatic aminoalkohols like Ph–CH(OH)–CH_2–NHMe and Ph–CH(OH)–CH(NHMe)–CH_3 [54].

A semi-conductor catalyst, Ag-doped TiO_2 allows the addition of both tri- and tetrachloromethane to 1-octene under UV-irradiation [55]. Hydrodechlorination takes place to a large extent under such conditions.

When interaction of an intermediate radical and Cl occurs in a coordination sphere of a metal (see Fig. 31), addition by a non-radical mechanism is potentially a way to stereoselective formation of polychloroalkanes. This is a potential way to obtain single enantiomers or diastereomers of PCA, or at least, a way to limit their number and improve purification procedure and characterization (Fig. 35).

However, no such synthesis of PCA has been attempted yet.

3.4 Telomerization of Small Olefins as a Way to Long-Chain PCA

Soon after the discovery of the Kharasch addition, it was found that small olefins may form telomers [56] (Fig. 36):

Telomer mixtures can be separated by rectification. Congeners containing up to 15 C atoms can be easily prepared this way [57]. Formation of telomers of general formula $C_nH_{2n-2}Cl_4$ with n = 20–50 was reported [58]. The Kharasch reaction with CCl_4 yields α,α,α,γ-tetrachloroalkanes, while telomerization opens a way to α,α,α,ω-tetrachloroalkanes. In the beginning free-radical initiators like azo-(bis)-diisobutyronitrile or benzoyl peroxide [57–60] were used; later, metal complexes became popular.

Chloroform can also act as telogen in the presence of metal carbonyl catalysts such as $Cr(CO)_6$, $Mo(CO)_6$, $V(CO)_6$, $Fe(CO)_5$ [33, 35, 61]. Carbonyls of Cr and Fe give good results. $Fe(CO)_5$ seems to be the best and most widely used. More complex telogens, 1,1,1-trichloroethane, and 1,1,1,3-trichloropropane were also investigated. For these compounds there is tendency to form shorter chain products [33, 34, 36]. Propylene, vinylchloride, and vinylidenechloride behave similarly to ethylene, but the telomerization extent is also smaller [33, 62] (Fig. 37).

An attempt to telomerize 1,1,1,5-tetrachloropentane (itself a 2:1 telomer of ethylene and tetrachloromethane) with ethylene, propylene, or vinylchloride in the presence of $Fe(CO)_5$ and isopropanol gave only 1:1 adducts (normal Kharasch addition) [33] (Fig. 38).

Telomerization of butadiene with 1,1,1-trichloroethane and 1,1,1,3-tetrachloropropane was achieved. Butadiene acts as conjugated diene in this reaction, but only 1:1 and 2:1 adducts were isolated [36] (Fig. 39).

An interesting transformation was observed upon attempted telomerization of allylchloride [34] (Fig. 40).

The intermediate 1,4,4,6-tetrachloro-2-hexyl radical may not only form 1,3,3,5,6-pentachlorohexane via abstraction of the Cl atom from 1,1,1,3-trichloropropane, but may eliminate the Cl radical as well with the formation of 4,4,6-trichlorohexene-1. This compound in turn gets involved into the Kharasch reaction

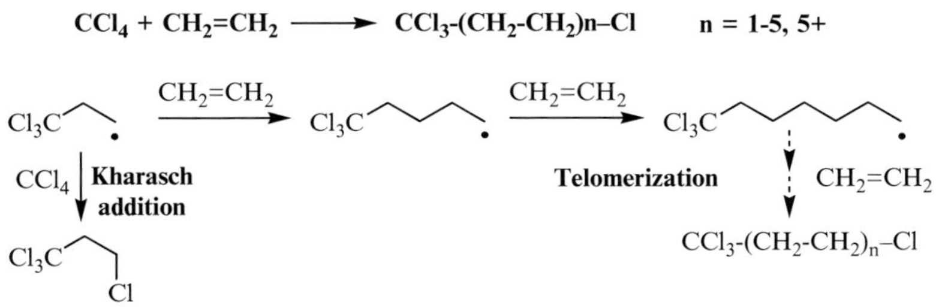

CCl4 + CH2=CH2 → CCl3-(CH2-CH2)n–Cl n = 1-5, 5+
CH2=CH2
CH2=CH2
Cl3C
CCl4
Kharasch addition
Telomerization
CCl3-(CH2-CH2)n–Cl
Cl

Fig. 36 Telomerization of ethylene and carbon tetrachloride and its relation to the Kharasch addition

$$HCCl_3 + CH_2{=}CH_2 \longrightarrow CCl_3\text{-}(CH_2\text{-}CH_2)_n\text{-}H \qquad n = 1\text{--}4,\ 4+$$

$$CH_3CCl_3 + CH_2{=}CH_2 \longrightarrow CH_3\text{-}CCl_2\text{-}(CH_2\text{-}CH_2)_n\text{-}Cl \qquad n = 1\text{--}3$$

$$ClCH_2CH_2CCl_3 + CH_2{=}CH_2 \longrightarrow ClCH_2CH_2\text{-}CCl_2\text{-}(CH_2\text{-}CH_2)_n\text{-}Cl$$

$$CH_3CCl_3 + CH_2{=}\overset{\overset{\textstyle CH_3}{\diagup}}{CH} \longrightarrow CH_3\text{-}CCl_2\text{-}(CH_2\text{-}CH)_n\text{-}Cl \qquad n = 1\text{--}3$$

$$ClCH_2CH_2CCl_3 + CH_2{=}CHCl \longrightarrow ClCH_2CH_2\text{-}CCl_2\text{-}(CH_2\text{-}CHCl)_n\text{-}Cl \qquad n = 1\text{--}3$$

$$ClCH_2CH_2CCl_3 + CH_2{=}CCl_2 \longrightarrow ClCH_2CH_2\text{-}CCl_2\text{-}(CH_2\text{-}CCl_2)_n\text{-}Cl \qquad n = 1,\ 2$$

Fig. 37 Telomerization of ethylene, propylene, vinylchloride and vinylidenechloride with chloroform, 1,1,1,-trichloroethane and 1,1,1,3-tetrachloropropane [33–36, 61, 62]

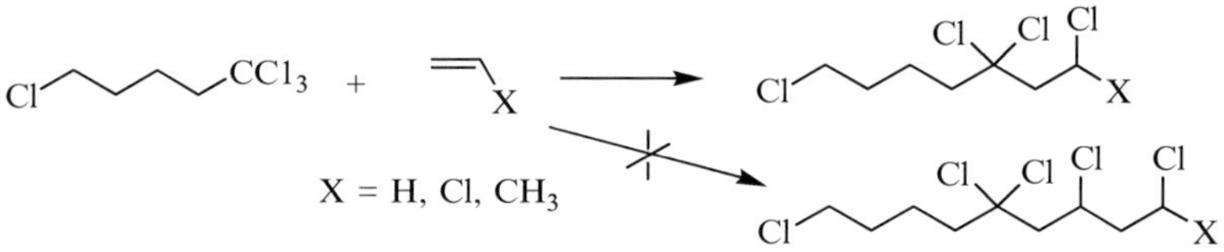

Fig. 38 Unsuccessful attempt of telomerization with 1,1,1,5-tetrachloropentane as telogen: only "normal" Kharasch adducts are formed [33]

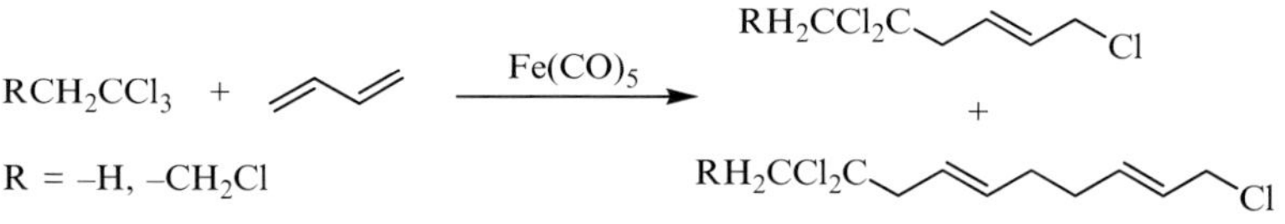

Fig. 39 Telomerization of butadiene with 1,1,1-trichloroethane and 1,1,1,3-tetrachloropropane [36]

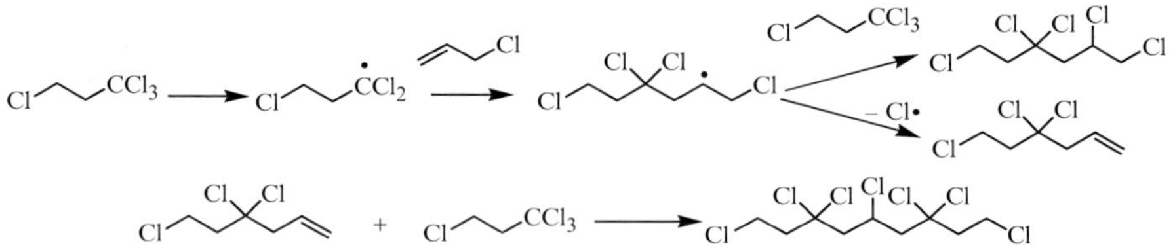

Fig. 40 Formation of 1,3,3,5,7,7,9-heptachlorononane in reaction of allylchloride with 1,1,1,3-trichloropropane [34]

with another molecule of 1,1,1,3-tetrachloropropane and yields 1,3,3,5,7,7,9-heptachlorononane.

Thus, telomerization not only opens a way to $\alpha,\alpha,\alpha,\omega$-tetrachloroalkanes up to 1,1,1,15-tetrachloropentadecane, but also to a variety of C_5–C_9 polychloroalkanes which can be used as CCl_3-components in the Kharasch synthesis of higher polychloroalkanes.

3.5 Modification of Polychloroalkanes: Hydrodechlorination, Coupling and Rearrangements

Terminal CCl_3 groups in polychloroalkanes can be easily converted to $CHCl_2$ groups by reaction with EtMgBr [63] or by reduction in the presence of $Fe(CO)_5$ and a hydrogen donor (dimethylformamide, hexamethylphosphotriamide, isopropanol) [64–67]. Hydrodechlorination with EtMgBr is accompanied by dimerization (Fig. 41).

Addition of $CoCl_2$ to reaction mixture increases the yield of 1,5,5,6,6,10-hexachlorodecane and decreases the yield of 1,1,5-trichloropentane. $Fe(CO)_5$ allows selective reduction of the CCl_3 group into $CHCl_2$ in the presence of CHCl and CH_2Cl groups in molecules with chain lengths of up to 11 carbon atoms [64, 65, 67, 68] (Fig. 42).

Reduction of 1,1,1-trichloroundecane also led to the formation of two rearrangement products – 1,1,5-trichlroundecane and 1,1,6-trichloroundecane [65] (Fig. 43). The two compounds were isolated.

The key step of the rearrangement is 1,5- or 1,6 migration of H in 1,1-dichloroundecyl-1 radical. During the reduction of 1,1,1,3-tetrachlorononane not only 1,5- and 1,6-H shifts were observed, but 1,7- and 1,8-H shifts as well [67] (Fig. 44).

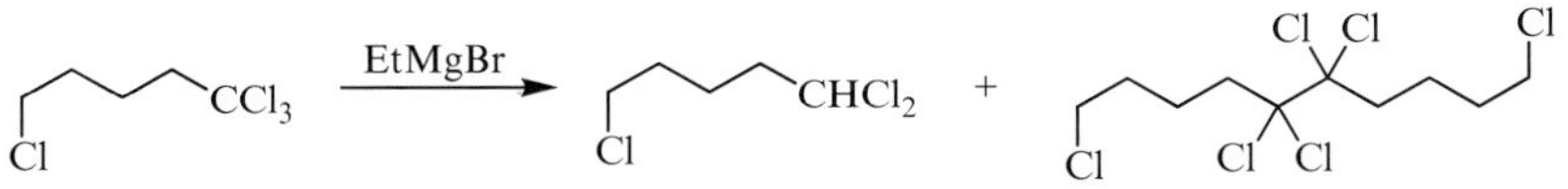

Fig. 41 Action of EtMgBr on PCA with terminal CCl_3-group: reduction and dimerization [63]

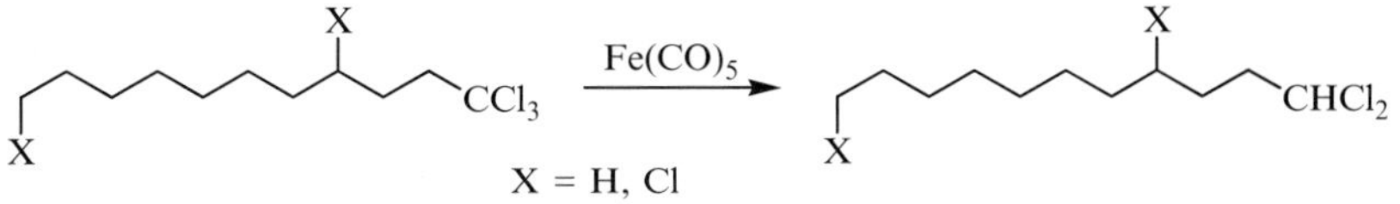

Fig. 42 Selective reduction of the CCl_3 group to $CHCl_2$ in presence of CHCl and CH_2Cl groups in a PCA molecule [64]

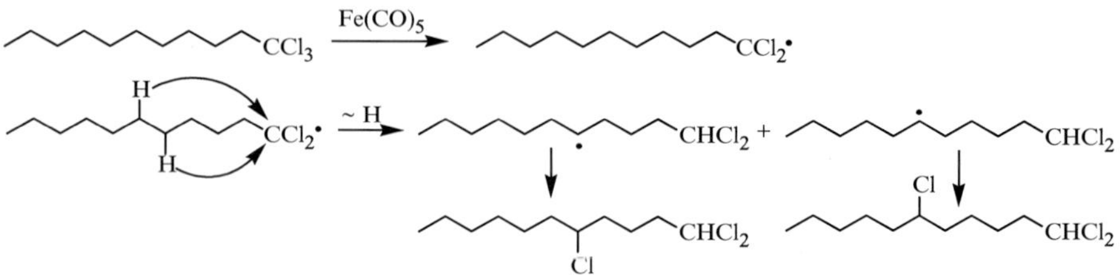

Fig. 43 Reduction of 1,1,1-trichloroundecane by $Fe(CO)_5$ accompanied by rearrangement: the way to 1,1,5-trichlroundecane and 1,1,6-trichloroundecane [65]

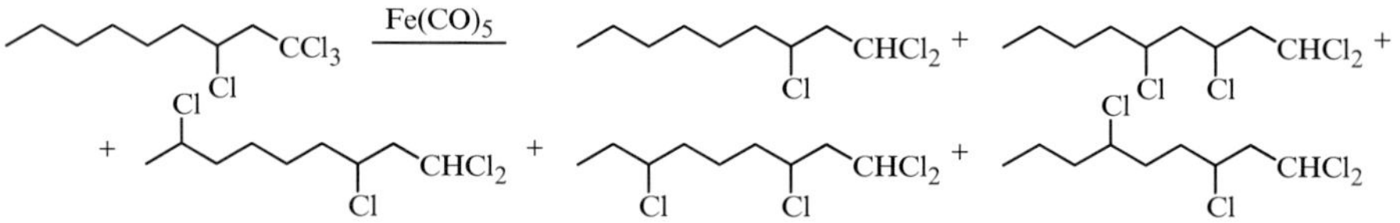

Fig. 44 Formation of 1,1,3,5-, 1,1,3,6-, 1,1,3,7- and 1,1,3,8-tetrachlorononanes as side products in the synthesis of 1,1,3-trichlorononane from 1,1,1,3-tetrachlorononane [67]

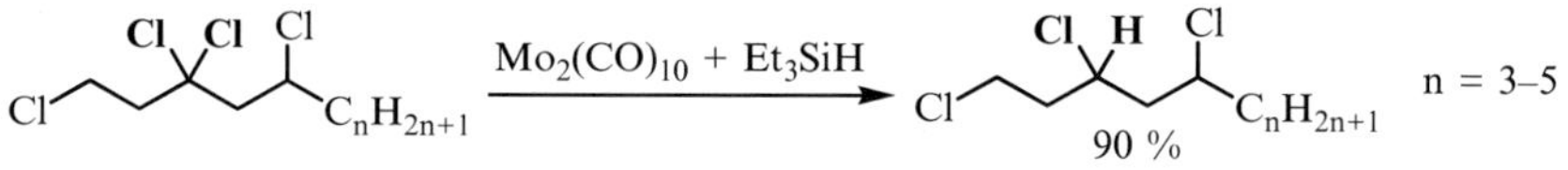

Fig. 45 Highly selective conversion of internal CCl_2 group in PCA into CHCl by reduction with the $Mo_2(CO)_{10}$ and Et_3SiH system [70]

Internal CCl_2 groups can also easily and selectively be reduced to CHCl [69, 70] on Fe, Mo, and Mn carbonyls in the presence of a hydrogen donor. The best yields (up to 90%) are reported for $Mo_2(CO)_{10}$ and Et_3SiH system [70] (Fig. 45).

Reactions using $Mo_2(CO)_{10}$ give very little rearrangement products, while the same reaction, catalyzed by $Fe(CO)_5$ gives products of 1,5- and 1,6-hydrogen shifts [71]. 1,3,5,7-tetrachlorononane and 1,3,5,8-tetrachlorononane were prepared this way (Fig. 46).

Hydrogenation of $\alpha,\alpha,\alpha,\omega$-tetrachloroalkanes with one terminal CH_2Cl group and one terminal CCl_3 group on palladium gives good yields of dimerization products [72, 73]; the major product is a hexachloroalkane. A side product, a tetrachloroalkene can be isolated and smoothly reduced to α,ω-dichloroalkane. Alternatively, the whole mixture can be hydrogenated with Zn and converted to α,ω-dichloroalkane with a very good yield (Fig. 47a).

Hydrolysis of telomeric $\alpha,\alpha,\alpha,\omega$-tetrachloroalkanes followed by electrolysis of ω-chloroalkancarboxylic acids is a good synthetic pathway to α,ω-dichloroalkanes up to 1,24-dichlorotetracosane [64] (Fig. 47b). α,ω-Dichloroalkanes themselves can be coupled elelctrolytically to α,ω-dichloroalkanes with a carbon chain of twice the

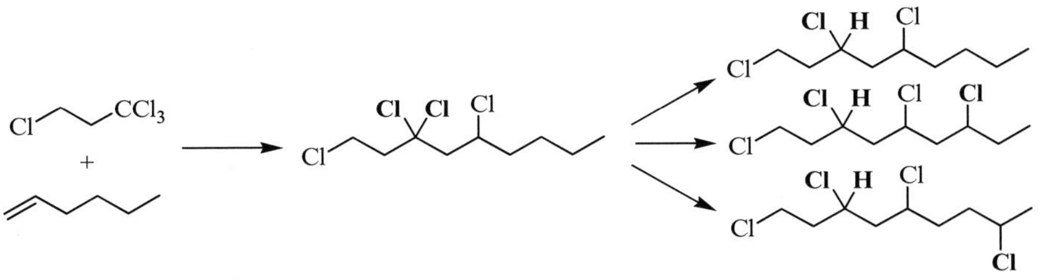

Fig. 46 Preparation of 1,3,5,7- and 1,3,5,8-tetrachlorononanes as a result of the rearrangement of intermediate radical during reduction of 1,3,3,5-tetrachlorononane [71]

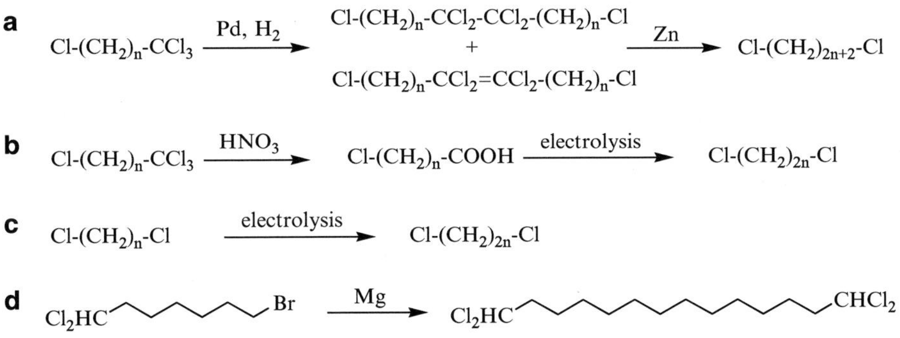

a $Cl\text{-}(CH_2)_n\text{-}CCl_3 \xrightarrow{Pd, H_2}$ $\begin{array}{c} Cl\text{-}(CH_2)_n\text{-}CCl_2\text{-}CCl_2\text{-}(CH_2)_n\text{-}Cl \\ + \\ Cl\text{-}(CH_2)_n\text{-}CCl_2\text{=}CCl_2\text{-}(CH_2)_n\text{-}Cl \end{array}$ $\xrightarrow{Zn}$ $Cl\text{-}(CH_2)_{2n+2}\text{-}Cl$

b $Cl\text{-}(CH_2)_n\text{-}CCl_3 \xrightarrow{HNO_3} Cl\text{-}(CH_2)_n\text{-}COOH \xrightarrow{electrolysis} Cl\text{-}(CH_2)_{2n}\text{-}Cl$

c $Cl\text{-}(CH_2)_n\text{-}Cl \xrightarrow{electrolysis} Cl\text{-}(CH_2)_{2n}\text{-}Cl$

d $Cl_2HC\diagdown\!\!\!\diagup\diagdown\!\!\!\diagup\text{-}Br \xrightarrow{Mg} Cl_2HC\diagdown\!\!\!\diagup\diagdown\!\!\!\diagup CHCl_2$

Fig. 47 Dimerization in the synthesis of polychloroalkanes: (**a**) reductive dimerization of terminal trichlorides followed by hydrogenation [72, 73]; (**b**) electrolytic dimerization of ω-chlorocarboxylic acids with loss of CO_2[50]; (**c**) electrolytic dimerization of α,ω-dichloroalkanes [50, 74]; (**d**) reductive coupling of α,α-dichloro-ω-bromoalkanes [75, 76]

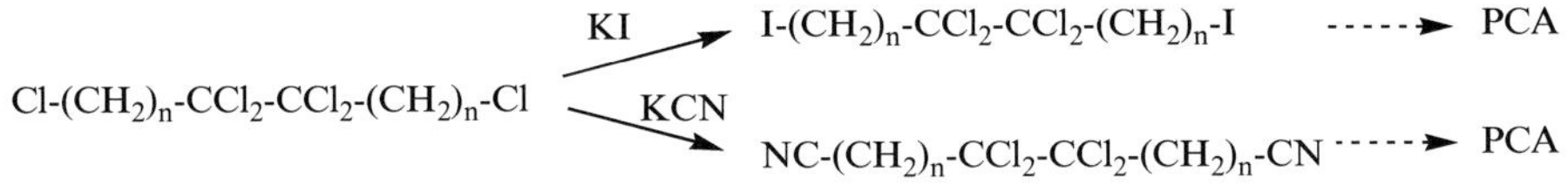

$Cl\text{-}(CH_2)_n\text{-}CCl_2\text{-}CCl_2\text{-}(CH_2)_n\text{-}Cl$ $\xrightarrow{KI}$ $I\text{-}(CH_2)_n\text{-}CCl_2\text{-}CCl_2\text{-}(CH_2)_n\text{-}I$ $\dashrightarrow$ PCA

$\xrightarrow{KCN}$ $NC\text{-}(CH_2)_n\text{-}CCl_2\text{-}CCl_2\text{-}(CH_2)_n\text{-}CN$ $\dashrightarrow$ PCA

Fig. 48 Nucleophilic substitution reactions of PCA: examples of intermediates for the synthesis of further PCA congeners [76]

length. [50, 74] (Fig. 47c). Telomerization of ethylene with bromodichloromethane followed by treatment with Mg leads to a coupling product with two terminal $CHCl_2$ groups: 1,1,14,14-tetrachlorotetradecane [75, 76] (Fig. 47d).

In reaction with magnesium, the $-CH_2Br$ terminal group is more reactive than $-CHCl_2$. It is worth mentioning that terminal C–Cl bonds are most reactive in nucleophilic substitutions. A number of useful derivatives can be prepared this way and used as intermediates for the preparation of other PCA congeners [76] (Fig. 48).

4   Conclusions

A sufficient number of PCA congeners are now available for the requirements of environmental chemistry.

Many of these congeners are in fact not single isomers, but mixtures of diastereomers or enantiomers.

At the moment, there is no clear need for the preparation of further individual compounds, as incentives based on toxicology, persistence, or even occurrence are still lacking.

There is a variety of established chemical methods that allows the preparation of literally any PCA congener, even in pure enantiomeric or diastereomeric form.

Appendix: Examples of Syntheses of Polychloroalkanes

A selection of methods taken from the original literature with experimental details for preparation of polychloroalkanes is collected in this Appendix. Methods are organized in order of increasing number of carbon atoms and then in order of increasing number of chlorine atoms. The original language is preserved. A complete list of syntheses is given in Table 2.

Examples of Synthetic Methods

1,1-Dichlorooctane (Fig. 17) and 1,1,X-trichlorooctanes (Fig. 20) [15]

1,115 g (5.350 mol) of phosphorus pentachloride and 500 g of benzene was placed into a 5-l., 3-neck flask. To the stirred slurry was added 600 g (4.7 mol) of octanal over a 5-h period at no more than 10°C. After standing overnight, ice and water were added slowly with cooling. The product was washed with sodium bicarbonate and water and then dried ($MgSO_4$). The product was distilled; the fraction boiling at 95–99°C was found to be 97% pure 1,1-dichlorooctane contaminated with octanal. Since the removal of octanal by sodium bisulfite washes was not successful, the product was redistilled to give 98.6% pure material.

1,1-X-Trichlorooctanes: to a 1-l. turbomixer, with an internal cooling coil and gas inlet at the bottom, was placed 499 g (2.58 mol) of 1,1-dichlorooctane. Nitrogen was bubbled through the mixture for 30 min. Chlorine was then passed with rapid stirring at about 25°C for 1 h at 258 mL/min with a GE sun lamp to initiate the reaction. Nitrogen was then bubbled through the mixture for 1 h and the reaction mixture washed with water, twice with 10% sodium bicarbonate, and finally with water. The mixture was dried ($MgSO_4$) and distilled. The fraction boiling at 117–130°C (15 mm) was used in the alkylation studies. The composition of this

Table 2 Literature references for the syntheses of individual PCA

PCA	References	PCA	References
1,1-Dichlorooctane	[13, 15, 65]	1,2,9,10-Tetrachlorodecane	[70]
1,2-Dichlorooctane	[13]	2,5,6,9-Tetrachlorodecane	[6]
1,4-Dichlorooctane	[13]	1,2,5,6,9-Pentachlorodecane	[6]
1,5-Dichlorooctane	[13]	1,1,1,3,9,10-Hexachlorodecane	[6]
1,8-Dichlorooctane	[50]	1,2,4,5,9,10-Hexachlorodecane	[8]
1,1,3-Trichlorooctane	[50]	1,2,5,6,9,10-Hexachlorodecane	[4, 6, 51]
1,1,5-Trichlorooctane	[65]	1,5,5,6,6,10,10-Hexachlorodecane	[63, 73]
1,1,6-Trichlorooctane	[65]	1,2,4,5,6,9,10-Heptachlorodecane	[8]
1,1,1,3-Tetrachlorooctane	[42]	1,2,5,5,6,9,10-Heptachlorodecane	[8]
1,1,1,3,6,8,8,8-Octachlorooctane	[26, 42]	1,1,1,3,8,10,10,10-Octachlorodecane	[6]
1,1-Dichlorononane	[64, 65]	2,3,4,5,6,7,8,9-Octachlorodecane	[8]
2,2-Dichlorononane	[14, 30]	1,2,3,4,5,6,7,8,9-Nonachlorodecane	[8]
1,1,1-Trichlorononane	[26, 37, 55]	1,1-Dichloroundecane	[64, 65]
1,1,5-Trichlorononane	[65]	1,1,5-Trichloroundecane	[65]
1,1,6-Trichlorononane	[65]	1,1,6-Trichloroundecane	[65]
1,1,9-Trichlorononane	[64, 66]	1,1,1,3-Tetrachloroundecane	[59]
1,3,3-Trichlorononane	[31]	1,2,10,11-Tetrachloroundecane	[4]
1,3,5-Trichlorononane	[70]	1,1,1,3,10,11-Hexachloroundecane	[6]
2,2,4-Trichlorononane	[30]	1,1,1,3,6,7,10,11-Heptachloroundecane	[6]
3,3,5-Trichlorononane	[9]	1,1,1,3,9,11,11,11-Octachloroundecane	[6]
1,1,1,3-Tetrachlorononane	[26, 39, 42, 46, 55, 59]	1,12-Dichlorododecane	[50, 73]
1,1,1,9-Tetrachlorononane	[56]	1,1,1,3,10,12,12,12-Octachloroundecane	[6]
1,3,3,5-Tetrachlorononane	[31]	1,1,1,3,6,7,10,12,12,12-Decachloroundecane	[6]
1,3,3,9-Tetrachlorononane	[33]	1,1,14,14-Tetrachlorotetradecane	[76]
1,3,5,7-Tetrachlorononane	[71]	1,2,13,14-Tetrachlorotetradecane	[4]
1,3,5,8-Tetrachlorononane	[71]	1,7,7,8,8,14,14-Hexachlorotetradecane	[73]
1,3,3,5,7,7,9-Heptachlorononane	[34]	1,16-Dichlorohexadecane	[50]
1,10-Dichlorodecane	[50, 73]	7,9,9,11-Tetrachloroheptadecane	[24, 39]
1,3,5-Trichlorodecane	[70]	1,9,9,10,10,18-Hexachlorooctadecane	[26]
1,2,5,6-Tetrachlorodecane	[6]	1,20-Dichloroeicosane	[64]
1,3,3,5-Tetrachlorodecane	[70]	1,24-Dichlorotetracosane	[64]

fraction was 1,1,1- (trace), 1,1,2- (2%), 1,1,3- (10.5%), 1,1,4- (16.1%), 1,1,5- (17.9%), 1,1,6- (21.0%), 1,1,7- (23.8%), and 1,1,8-trichlorooctane (10.3%).

After the alkylation reaction was complete, the 1,1,8 isomer and a mixture of the 1,1,2 and 1,1,3 isomers were trapped from a 10 ft × 0.25 in. 20% Carbowax on Chromosorb W column.

The NMR of the 1,1,8 isomer showed a triplet at 5.56 (CHCl), a triplet at 3.38 (CH_2Cl), a crude quartet at 2.16 (CH_2CCl_2), and methylene protons at 1.37 ppm.

The mixture of 1,1,2 and 1,1,3 isomers – the latter being predominant – showed a pair of overlapping doublets at 5.8, a smaller doublet at 5.71, a complex peak at 3.96, a pair of overlapping doublets at 2.42, methylene protons at 1.3, and terminal methyl at 0.89 ppm. The mixture was then subjected to spin-spin decoupling treatment. Irradiation at 3.9 ppm (a) collapsed the 5.8 ppm band to a triplet, (b) collapsed the 5.71 ppm doublet to a singlet, thereby giving a good indication that the minor constituent was the 1,1,2 isomer, and (c) collapsed the 2.42 ppm doublet pair to one doublet. Irradiation at 2.5 ppm reduced the 5.8 ppm doublets to a singlet and the complex group at 3.96 ppm to a crude triplet. Irradiation at 5.8 ppm did not affect the 3.96 ppm group but reduced the bands at 2.42 ppm to a doublet. C-3 of the 1,1,3 isomer is asymmetric; hence the protons on C-2 are not magnetically equivalent. The assignment of the bands at 5.8 (Cl_2CH), 3.96 ($CHCl$), and 2.42 ppm (Cl_2CCH_2CCl), and the behavior in the spin–spin decoupling treatment are consistent with that for the 1,1,3 isomer.

1,1,1-Trichlorooctane and 1,1,1,X-tetrachlorooctanes (Fig. 20) [15]

1,1,1-Trichlorooctane: Into a 1-l., stainless steel, stirred autoclave was placed 186 g (1.90 mol) of 1-heptene, 900 g (6.56 mol) of chloroform, and 4.0 g of benzoyl peroxide. The sealed bomb was heated at 80°C for 4 h and a further 8 g of benzoyl peroxide added. The bomb was then heated for another 6 h at 90°C. This procedure was repeated; the two products were combined. The unreacted chloroform and 1-heptene were removed by distillation, and the bulk of the product distilled at 84–86°C (6 mm). More careful fractionation of the 240 g of product so obtained gave a product of 98.5% purity boiling at 85°C (6 mm).

1,1,1,X-Tetrachlorooctanes: The chlorination of 1,1,1-trichlorooctane used the previously described procedure (See 5.1.1). The fraction boiling at 115–123°C (6 mm) was used in the alkylation studies. The composition of this fraction was 1,1,1,3- (3.2%), 1,1,1,4- (19.7%), 1,1,1,5- (22.9%), 1,1,1,6- and 1,1,1,7- (46.7%), and 1,1,1,8-tetrachlorooctane (6.3%). The impurities plus the 1,1,1,2 isomer constituted 1.2% of the mixture. Concentration of various isomers, either from distillation cuts before reaction or by trapping the unreactive isomers after reaction, permitted spectroscopic methods to substantiate the assignments. Further, the 1,1,1,3 isomer was synthesized by the addition of carbon tetrachloride to 1-heptene [b.p. 105–106°C (6 mm)].

The NMR of the 1,1,1,3 isomer showed peaks at 4.2 (m, $CHCl$), 3.18 (eight line distinctive multiplet), 1.82 (m, $CClCH_2$), 1.4 (methylene protons), and 0.99 ppm (terminal methyl). The most distinctive feature from the NMR of the 1,1,1,6 isomer was the triplet methyl group at 1.14 ppm. The doublet methyl group of the 1,1,1,7 isomer was observed in the NMR at 1.6 ppm. The NMR of the 1,1,1,8 isomer showed a band at 3.4 ppm (t, $ClCH_2$) and was distinctive because of the absence of the terminal methyl group.

1,1,5-Trichlorooctanes and 1,1,6-Trichlorooctanes, 1,1,5-Trichlorononanes and 1,1,6-Trichlorononanes, 1,1,5-Trichloroundecanes and 1,1,6-Trichloroundecanes (Fig. 43) [65]

A mixture of 4.35 g (20 mmole) of 1,1,1-Trichlorooctane, 9.6 g (160 mmole) of *iso*-C_3H_7OH, and 0.59 g (3 mmole) of $Fe(CO)_5$ was heated at 120°C for 3 h. The reaction mixtures from five experiments were combined and distilled with collection of the fraction that boiled up to 80°C (*iso*-C_3H_7OH and its transformation products). The residue was washed with 10% HCl and water and distilled in vacuo to give 12.4 g (65%) of 1,1-dichlorooctane with b.p. 85–87°C (15 mm). According to GLC data the fraction with b.p. 97–128°C (10 mm) (4.2 g) contained two substances. Preparative GLC yielded 1,1,5- and 1,1,6-trichlorooctanes in the form of 95% enriched fractions containing the second isomer.

Reduction of 1,1,1-Trichlorononane in the Presence of $Mn_2(CO)_{10}$. A mixture of 20 mmole of 1,1,1-Trichlorononane, 160 mmole of *iso*-C_3H_7OH, and 2 mmole of $Mn_2(CO)_{10}$ was heated at 120°C for 3 h. The reaction mixtures from three experiments were combined and worked up to give 9.4 g (68%) of 1,1-dichlorononane with b.p. 102–104°C (15 mm). The undistilled residue (1.6 g) contained 1,1,5- and 1,1,6-trichlorononanes, which were identified by GLC with reference to samples obtained by an independent method.

Reduction of 1,1,1-Trichloroundecane in the Presence of $Mo(CO)_6$. A mixture of 20 mmole of 1,1,1-Trichloroundecane, 160 mmole of *iso*-C_3H_7OH, and 1 mmole of $Mo(CO)_6$ was heated at 140°C for 3 h. The reaction mixtures from five experiments were combined and worked up. Distillation gave 16.6 g (74%) of 1,1-dichloroundecane with b.p. 101–102°C (2 mm) and a fraction containing, according to GLC data, 3.2 g (16%) of a mixture of 1,1,5- and 1,1,6-trichloroundecanes, which were identified by GLC with reference to samples obtained by an independent method [64].

1,1,1,3-Tetrachlorooctane, 1,1,1,3-Tetrachlorononane, 1,1,1,3,6,8,8,8-Octachlorooctane (Fig. 30) [42]

All reactions and preparative manipulations were carried under nitrogen atmosphere, using Schlenck techniques. Acetonitrile was distilled from P_2O_5 prior to use. Propionitrile and benzonitrile, reagent grade, were used without purification. In a typical reaction, the catalyst, olefin substrate and the polyhalide, in a molar ratio of 0.05:1:6.6, were refluxed in the solvent under dry nitrogen atmosphere for the time specified; [olefin substrate] = 0.7 M.

Isolation of the product was conducted by distilling the solvent and the excess polyhalide, followed by precipitation of the catalyst (oxidized) with pet ether followed by passing the resulting clear solution through a short silica column. The crude product was then purified by vacuum distillation.

1,1,1,3-Tetrachlorononane (24h), 85%, b.p. 112 8°C/1 mmHg. ^{1}H-NMR: δ = 0.9 (t, 3H), 1.4 (m, 8H), 1.9 (m, 2H), 3.2 (ddd, 2H), 4.2 (m, 1H). MS: m/z (%) = 223(19), 193(29), 185(35), 157(100), 149(29), 132(58), 121(94), 109(73).

1,1,1,3-Tetrachlorooctane (23h), 80%, b.p. = 54°C/0.1 mmHg. ^{1}H-NMR: δ = 0.9 (t, 3H), 1.5 (m, 6H), 1.9 (m, 2H), 3.2 (ddd, 2H), 4.2 (m, 1H). MS: m/z (%) = 223(7), 187(11), 179(23), 143(54), 107(51), 97(43), 82(58), 69(67), 55(97), 41(100).

1,1,1,3,6,8,8,8-Octachlorooctane (22h), 22%, b.p. = 114–120°C/0.3 mmHg. ^{1}H-NMR: δ = 1.6–2.4 (m, 4H), 3.2 (dddd, 4H), 4.3 (m, 2H). MS: m/z (%) = 317(50), 281(33), 245(50), 221(30), 183(28), 159(31), 143(55), 123(44), 109(100), 87(33), 75(58), 61(23), 43(58).

1,1,1,3,6,8,8,8-Octachlorooctane (Fig. 24) [26]

A mixture of biallyl (41 g., 0.5 mole), carbon tetrachloride (318 g., 2.07 moles), and acetyl peroxide (3.4 g., 0.029 mole) was heated to reflux under an excess pressure of 15 cm. of mercury. After 5 h heating, the boiling point had risen from 85°C to 96°C. In the −80°C cold trap, a small amount of methyl chloride was collected. The excess carbon tetrachloride was distilled from the reaction mixture, and the residue was fractionated under reduced pressure. A fraction boiling at 50–60°C (0.4 mm.) was collected. Upon distillation this material yielded 31g. of tetrachloroheptene; b.p. 57–59°C (0.4 mm.).

The residual portion of the reaction mixture (85 g.) was distilled in a molecular still. Successive portions of the distillate (60 g.) showed a progressively decreasing chlorine content: from 69.2 to 63.8%. On standing, fine white needles were formed by partial crystallization of the distillate. These were washed quickly with a little methyl alcohol and dried. This material (m. p. 72–74°C) appeared to be octachloro-octane.

1,1-Dichlorononane (Fig. 42) [64]

The reaction product from 1,1,1-Trichlorononane (10.2 g, 50 mmol) was distilled in vacuum. Three fractions were obtained: (1) b.p. 77–80°C (10 mm) 2.5 g; (2) b.p. 62–70°C (1 mm) 2.4 g; (3) b.p. 78–80°C (1 mm) 5 g.

1,1-Dichlorononane was isolated from fraction 1 by a repeated distillation and had b.p. 59°C (2 mm). ^{13}C NMR spectrum (δ, ppm): 73.5 (C-1), 44.0 (C-2), 28.8 (C-3), 26.2 (C-4), 29.6 (C-5), 29.4 (C-6), 32.1 (C-7), 22.8 (C-8), 14.1 (C-9).

2,2-Dichlorononane (Fig. 17, Reaction 2) [14]

Chlorine was bubbled by stirring in dichloromethane (40 mL) containing 2-nonan-one oxime (2 g) and aluminum trichloride (1 g). The mixture was poured over ice (100 g) and then extracted with ether (100 mL). The organic layer was washed with

5% hydrochloric acid, sodium hydrogen carbonate solution, and brine and then dried over magnesium sulfate. 2,2-Dichlorononane: yield 82%, b.p. = 59°C (30.8 Torr), ^{1}H NMR 2.05 ppm (COCH$_2$), 2.1 ppm (CH$_3$). ^{13}C NMR 37.2(C-1), 90.6(C-2), 49.7(C-3), 25.6 (C-4), 28.8–28.4(C-5, C-6), 31.6(C-7), 22.6(C-8), 13.9(C-9).

1,1,1-Trichlorononane, A (Fig. 37) [37]

The radical telomerization of 20 g of 1-octene (0.178 mol) and 213 g of chloroform (1.78 mol) was initiated by 0.71 g of di(p-*tert*-butylcyclohexyl)percarbonate (1.78 mmol). The reaction was performed under argon at 60°C for 10 h. A colorless oil was distilled: $CCl_3CH_2(CH_2)_6CH_3$, (yield 52%; b.p. = 50°C/2 mm Hg). ^{1}H-NMR (CDCl$_3$): δ = 0.9 (t; 3H; CH$_3$), 1.3 (m; 10H; C$_5$H$_{10}$CH$_3$), 1.8 (m; 2H; CCl$_3$CH$_2$CH$_2$), 2.65 (2H; m; CCl$_3$CH$_2$). ^{13}C-NMR (dmf-d_7): δ = 14.4 (CH$_3$, 1C), 23.1–32.5 (C$_6$H$_{12}$CH$_3$, 6C), 55.4 (CCl$_3$CH$_2$, 1C), 101.3 (CCl$_3$, 1C).

1,1,1-Trichlorononane, B (Fig. 23) [26]

Octene-1 (28 g., 0.25 mole), chloroform (120 g., 1 mole), and benzoyl peroxide (0.5 g.) were mixed and heated under 20 cm excess pressure for ten hours during which time the boiling point of the mixture rose from 80 to 92°C. After 4 h, an additional amount of benzoyl peroxide (1.0 g., total 0.006 mole) was added. On distillation of the reaction mixture, about 1.5 g. of unchanged octene was recovered; on further distillation, a fraction boiling at 65–75′ (0.1 mm.) was obtained. When redistilled, this material gave a product which was shown to be 1,1,1-trichlorononane, 13 g. (22%); b.p. 65–70°C (0.5 mm.).

1,1,9-Trichlorononane, A (Fig. 42) [64]

1,1,1,9-Tetrachlorononane was reduced with a system consisting of Fe(CO)$_5$ (15 mole %) and HMPA (100 mole %) at 140°C for 3 h in a stream of nitrogen. The reaction mixture was processed, distilled, and from the fraction of b.p. 78–80°C (10 mm) containing 95% of 1,1,9-trichlorononane and 5% 1,1,1,9-tetrachlorononane, 1,1,9-trichlorononane was isolated by preparative GLC.

^{13}C NMR spectrum (δ, ppm): 73.7 (C-1), 43.7 (C-2), 25.9 (C-3), 29.3 (C-4), 28.8 (C-5), 28.5 (C-6), 26.9 (C-7), 32.7 (C-8), 44.8 (C-9).

1,1,9-Trichlorononane, B (Fig. 41) [66]

A solution of 0.77 g of Fe(CO)$_5$ in 20 g of n-C$_4$H$_9$SH was added with stirring to 22 g of 1,1,1,7-tetrachloroheptane heated to 145°C at such a rate that the temperature in the reaction mixture was 140–145°C. The HCl liberated was collected in water, and was determined quantitatively by the titration of an aliquot part with 0.1 N NaOH solution at the end of the reaction.

The yield of HCl was 73% of theoretical. The di-n-butyl disulfide formed was isolated from the reaction mixture by distillation through a column with a yield of 83% of theoretical. By distillation from a Favorskii flask, the residue yielded 14.1 g (75%) of 1,1,7-trichloroheptane contaminated with a small amount of dibutyl disulfide. After redistillation, the 1,1,7-trichloroheptane was obtained in the pure state according to the results of GLC and analysis.

1,1,9-trichlorononane was obtained similarly; 71%, b.p. 123°C (3 mm Hg).

1,3,3-Trichlorononane and 1,3,3,5-Tetrachlorononane (Fig. 25) [31]

Reaction of 1,1,1,3-Tetrachloropropane with 1-Hexene.

All the experiments were run and worked up by the general procedure described in [31]. Fractional distillation of the reaction mixture from the experiment run at 105°C in the presence of hexamethylphosphotriamide + N,N-dichloro-p-chloro-benzenesulfonamide gave 22.5 mmoles of 1,3,3,5-tetrachlorononane, b.p. 126°C (3m m), PMR spectrum (δ, ppm): 0.70–2.07m (C_4H_9, 9H), 2.72m (4H, $2CH_2$), 3.82t (2H, CH_2Cl), 4.25m (1H, CHCl).

From the reaction mixture by preparative GLC we isolated 1,3,3-trichlorononane, b.p. 85–88°C (1 mm); PMR spectrum (δ, ppm): 0.65–1.90m (11H, C_5H_{11}), 2.17m (2H, CH_2), 2.60t (2H, CH_2CCl_2), 3.84t (2H, CH_2Cl).

2,2,4-Trichlorononane and 2,2-dichlorononane (Fig. 25) [30]

A mixture of 5 g olefin, 33.4 g of 1,1,1-Trichloroethane, 17 g of $HSiEt_3$, and 1.5 g of *tert*-butylperoxide (TBP) was heated at 130–140°C. Gas–liquid chromatographic analysis of the reaction products indicated the formation of 3.3 g (33%) 2,2-dichlorononane and 1.2 g (10%) 2,2,4-trichlorononane. The mixture was fractio-nated after washing with dilute hydrochloric acid, water, extraction, and drying. The fraction with b.p. 106–111°C (16 mm) (7.6 g) was subjected to preparative gas–liquid chromatography to give 2,2-dichlorononane. PMR spectrum (δ, ppm): 0.84 m, 1.2 m (CH_3, $6CH_2$, 15H), 2.0 s (CH_3CCl_2, 3H).

The fraction with b.p. 112–116°C (16 mm) (4.6 g) was subjected to chromatogra-phy to yield 2,2,4-trichlorononane. PMR spectrum (δ, ppm), 2.18 s (CH_3CCl_2, 3H), 2.6 d (CCl_2CH_2, 2H), 4.0 m (CHCl, 1H), and 0.94 m and 0.9 m ($4CH_2$, CH_3, 11H).

3,3,5-Trichlorononane (Fig. 25) [27]

A mixture of 0.39 M of $CCl_3C_2H_5$, 0.33 M of 1-hexene, 0.78 M of *iso*-C_3H_7OH, (abs.), and 6 mM of $Fe(CO)_5$ was heated for 5 h in an autoclave at 135°C. After distilling off fraction I, 47.8 g, b.p. 36–86°C, and washing out the iron salts, fraction

II was obtained by distillation with b.p. 67–84°C (2 mm), 32.4 g, and a residue of 9.5 g which was not investigated. From the GLC results, fraction II contained 0.061 M of $CH_3CH=CClCH_2CHClC_4H_9$ and 0.082 M of $C_2H_5CCl_2CH_2–CHClC_4H_9$.

By repeated distillation of fraction II we isolated 9 g of 3,5-Dichloro-2-nonene (11.8% of theor.) From fraction II we isolated 16.9 g (18.8%) of 3,3,5-Trichloro-nonane.

1,1,1,3-Tetrachlorononane (Fig. 23) and 7,9,9,11-tetrachloroheptadecane [26]

Reaction of Carbon Tetrachloride with Octene-1: Octene-1 (b.p. 121.2°C (750 mm.); (37 g., 0.33 mole), carbon tetrachloride (154 g., 1.0 mole) and benzoyl peroxide (5 g., 0.02 mole) were heated together under an excess pressure of 15 cm. of mercury. Carbon dioxide was steadily evolved for about 4 h, during which time the boiling point of the reaction mixture rose from 90 to 105°C. The excess of carbon tetrachloride was then removed by distillation, and the residue was distilled in vacuo. The forerun contained a small amount of a white solid material. A fraction (72 g) boiling at 75–85°C (0.05 mm.) was collected and redistilled. The yield of redistilled product was 66 g. (75%); b.p. 78–79°C (0.1 mm.). This material was 1,1,1,3-tetrachlorononane.

The residue (12.5 g.) was distilled in a molecular still. Three fractions were taken; these showed a progressive decrease in chlorine content from 46.8 to 41.0%. A compound consisting of two moles of octene and one mole of carbon tetrachloride contains 37.6% chlorine. Therefore, the high-boiling material is probably a mixture of $C_9H_{16}Cl_4$ and $C_{17}H_{32}Cl_4$.

When acetyl peroxide instead of benzoyl peroxide was used to initiate the reaction of carbon tetrachloride with octene-1, the results were similar. The yield of 1,1,1,3-tetrachlorononane obtained was 85% of the calculated amount.

The Reaction of Carbon Tetrachloride with Octene-1 in Ultraviolet Light: Carbon tetrachloride (182.1 g., 1.18 mole) and octene-1 (39.9 g., 0.36 mole) were mixed in a quartz reflux apparatus. The mixture was held at its boiling point and irradiated with a 500-watt ultraviolet lamp for a period of 4 h. After the unchanged carbon tetrachloride and octene-1 had been removed by distillation, 1,1,1,3-tetra-chlorononane (b.p. 72–75°C (0.1 mm.) distilled. A residue (3.9 g.) remained in the distilling flask.

1,1,1,9-Tetrachlorononane (Fig. 36) [56]

A stainless steel-lined tubular pressure reactor having an internal volume of about 350 mL and equipped with a thermocouple well and gas inlet was charged with 210 g. (1.36 moles) of freshly distilled carbon tetrachloride, 35 g. of water, and 0.47 g. (0.00194 mole) of benzoyl peroxide ("Lucidol"). The reactor was evacu-ated, pressured to 500 lb./sq. in. with ethylene, and placed horizontally in a shaking box equipped with a heater. When the temperature of the reaction mixture was raised to 70°, the pressure was increased to 1,400 lb./sq. in. by injection of ethylene, and the heating was continued. The reaction mixture was maintained at 95°C, and the pressure in the range 1,200–1,400 lb./sq. in. by injection of additional ethylene as required, for 5 h. The reaction product was then removed from the cooled reactor, separated from the water, and dried over anhydrous magnesium sulfate. After removal of the unreacted carbon tetrachloride by distillation, a preliminary fractional distillation gave the following results:

b.p. <90°C, 7.2%(C_3)
b.p. 90–115°C, 59.7%(C_5)
b.p. 115–145°C, 22.2%(C_7)
b.p. 145–175°C, 7.7%(C_9)
residue 3.2%(>C_9)

The pure compounds can be obtained from these cuts by redistillation.

1,3,3,5,7,7,9-Heptachlorononane (Fig. 40) [34]

A mixture of 2.8 moles of 1,1,1,3-Tetrachloropropane, 2.8 moles of allyl chloride, 350 mL of i-C_3H_7OH, and 49 mmoles of $Fe(CO)_5$ was heated in an 0.5-L steel autoclave for 2 h at 135°C. The reaction products were washed with 15% HCl solution, extracted with $CHCl_3$, dried over $MgSO_4$, and fractionally distilled. The following fractions were obtained: (I) b.p. 51–65°C (30 mm), 206.8 g; (II) b.p. 78–100°C (30 mm), 11 g; (III) b.p. 42–56°C (3 mm), 15.2 g; (IV) b.p. 79–93°C (1 mm), 29.7 g; (V) b.p. 92–101°C (0.9 mm), 172 g. The residue weighed 54.5 g. Based on the GLC analysis, fraction I contained 1.2 moles of pure 1,1,1,3-Tetrachloropropane (57%conversion). Fractions II and III respectively contained 90 and 70% of 4,4,6-trichloro-1-hexene. The pure 4,4,6-trichloro-1-hexene was isolated from fraction II by preparative GLC. Fractions IV and V respectively contained 35 and 80% of 1,2,4,4,6-pentachlorohexane (36% when based on reacted 1,1,1,3-Tetrachloropropane). GLC analysis on two phases of different polarity disclosed that fraction IV also contained 1,3,3,5-tetrachlorohexane. Fraction V was treated with conc. H_2SO_4 to remove the by-products that were formed by the alkoxylation of the allylic Cl atom, washed with water, dried over $MgSO_4$, and repeatedly distilled to give the pure 1,2,4,4,6-pentachlorohexane. The residue (54.5 g) was extracted with petroleum ether. Distillation of the extract gave a fraction with b.p. 150–153°C (0.8 mm), which, based on the GLC analysis and NMR spectral data, contains 95% of 1,3,3,5,7,7,9-heptachlorononane ($ClCH_2CH_2CCl_2CH_2)_2CHCl$.

1,10-Dichlorodecane (Fig. 47a) [73]

A mixture of 100 g of 1,5,5,6,6,10-hexachlorodecane, 100 g of diethylamine, 200 mL of ethanol, and 12 g of Raney nickel was placed in a 1-L rotating autoclave. Hydrogen was passed into the autoclave until the pressure attained 100 atm.

The reaction was carried out at 50–55°C, hydrogen being passed in as necessary so that the pressure was maintained at 100–120 atm. After 8 h, reaction ceased. The products obtained were 41 g (68%) of 1,10-dichlorodecane (b.p. 105–106°C (1.5 mm) and 15 g (19%) of 1,5,6,10-tetrachlorodecene-5.

1,2,5,6-Tetrachlorodecane and 1,1,1,3,10,12,12,12-Octachlorododecene (Figs. 4 and 5) [6]

Hydrogenation of the compounds 5,6,9,10-tetrachlorodecenes-1, 1,2,5,9,10-penta-chlorodecene-5, 1,1,1,3,10,12,12,12-octachlorododecenes-6: 0.5 g of 5,6,9, 10-tetrachlorodecenes-1 (or 1,2,5,9,10-pentachlorodecene-5, 1,1,1,3,10,12,12, 12-octachlorododecenes-6) was dissolved in 50 mL of ethylacetate in a round-bottom flask with two necks, and 50 mg Pd on activated carbon was added. Hydrogenation of the magnetically stirred sample was carried out under about 5 psi H_2 of overpressure. After completion of the reaction, the solution was filtered, the solvent rotary evaporated and redissolved in petroleum ether. Contrary to 5,6,9,10-tetrachlorodecenes-1, and 1,2,5,9,10-pentachlorodecene-5, the hydrogenation products of 1,1,1,3,10,12,12,12-octachlorododecenes-6 were identical, and could be isolated only in about 50% purity after threefold chromatography on silica gel column (100 × 2.6 cm). Therefore, these samples were further chromatographed on a GPC column with hexane–dichloromethane (1:1). Finally, purification on silica gel was again necessary to obtain the product with a purity of more than 90%.

1,2,9,10-Tetrachlorodecane (Fig. 2) [3]

1,2,9,10-Tetrachlorodecane was synthesized by chlorine addition to 1,9-decadiene using a variation of the procedure reported by Tomy [4]. Approximately 10 mL of 0.05 M NaOH was layered over 40 mL of dichloromethane in a round bottom flask. Chlorine gas was gently bubbled through the DCM layer for several minutes before introduction of 1,9-decadiene into the lower DCM layer by pipette. After a brief reaction period, the layers were separated and the DCM layer was analyzed by GC/ MS. The total ion chromatogram produced one dominant peak with a retention time of 13.7 min and an EI mass spectrum confirming the identity of the product as tetrachlorodecane. The very small amounts of higher chlorinated decane isomers (with chlorine number > 4) evident in the chromatogram indicated that free radical substitution reactions were minimal.

1,3,3,5-Tetrachlorodecane, 1,3,3,5-Tetrachlorononane and 1,3,5-Trichlorodecane (Fig. 25) [70]

1,3,3,5-Tetrachlorodecane was obtained similarly to 1,3,3,5-tetrachlorononane from 250 mmoles of 1-heptene, 750 mmoles of 1,1,1,3-tetrachloropropane, 25 mmoles of Fe

(CO)$_5$, and 100 mmoles of DMF at 105°C in the course of 3 h. The yield of 1,3,3,5-Tetrachlorodecane was 52%, b.p. 117°C (2 mm). ^{13}C NMR spectrum (δ, ppm): 38.8 (C-1) 49.8 (C-2), 89 9 (C-3), 56.5 (C-4), 57.8 (C-5), 39.4 (C-6), 31.0 (C-7), 25.5 (C-8), 22.3 (C-9), 13.8 (C-10).

Reduction of 1,3,3,5-tetrachloroalkanes was carried out in sealed glass ampoules (140°C, 3 h, with rotary stirring). 1,3,5-Trichlorodecane was obtained from 25 mmoles of 1,3,3,5-Tetrachlorodecane. The mixture was passed through a layer (30 mm) of silica gel L 100/160, and the silica gel was washed with 20 mL of CCl$_4$. After distillation of low-boiling products, the mixture was distilled in vacuo; 1,3,5-Trichlorodecane was isolated, yield 84%, b.p. 114°C (2 mm), ^{13}C NMR spectrum (δ, ppm): 40.3 (C-1), 40.9 (C-2), 56.4 (C-3), 46.9 (C-4), 58.8 (C-5), 37.9 (C-6), 31.1 (C-7), 25.7 (C-8), 22.5 (C-9), 13.9 (C-10).

2,5,6,9-Tetrachlorodecane and 1,2,5,6,9-Pentachlorodecane (Fig. 4) [6]

0.5 g of 5,6-dichlorodecadienes-1,9 or 5,6,9,10-tetrachlorodecenes-1 and 100 µl SnCl$_4$ were dissolved in 15 mL of water-free dioxan in a 20 mL headspace vial. HCl gas generated from conc. hydrochloric acid by heating in a round-bottom flask and dried over CaCl$_2$ was introduced into the dioxan solution up to saturation. After this, the sealed vial was kept in an oven at 60°C for about 2 days. The course of the reaction was followed gas chromatographically by injection of samples taken every 16 h with a GC syringe. The products were extracted from dioxan with petroleum ether after the addition of water. For the purification, a silica gel column (100 × 2.6 cm) with petroleum ether as eluent was used.

1,1,1,3,9,10-Hexachlorodecane, 1,1,1,3,6,7,10,11-Octachloroundecane, 1,1,1,3,6,7,10,12,12,12-Octachlorododecane (Fig. 5) [6]

Chlorination of the compounds 9,11,11,11-tetrachloroundecadienes-1,5 and 1,1,1 ,3,10,12,12,12-octachlorodecenes-6, 8,10,10,10-tetrachlorodecene-1 and 9,11,11, 11-tetrachloroundecene-1. 300 mg of the compound was dissolved in 25 mL of CCl$_4$, to which 250 mg chlorine in 5 mL CCl$_4$ was added under exclusion of light. From 9,11,11,11-tetrachloroundecadienes-1,5, first hexachloroundecenes were formed, which were not isolated and the reaction was continued to 1,1,1, 3,6,7,10,11-octachloroundecane.

Chlorination of 1,1,1,3,10,12,12,12-octachlorodecenes-6 yielded 1,1,1,3,6,7,10, 12,12,12-octachlorododecane, and that of 8,10,10,10-tetrachlorodecene-1 yielded the product 1,1,1,3,9,10-hexachlorodecane. By the same method, compound 1,1,1,3,10,11-hexachloroundecane was obtained from 9,11,11,11-tetrachloroundecene-1. The purification of all end products was carried out on silica gel column with petroleum ether as described above.

1,2,5,6,9,10-Hexachlorodecane (Fig. 4) [6]

Chlorination of 1,5,9-decatriene: 1.75 g chlorine dissolved in 100 mL of CCl_4 was slowly dropped into an intensively magnetically stirred solution of 2.5 g 1,5,9-decatriene in 50 mL of CCl_4 at 0°C. Another reaction charge was carried out under the same reaction conditions with 2.5 g 1,5,9-decatriene and 3.5 g chlorine. After completion of the reaction of chlorine, the two solutions were pooled and the solvent was rotary evaporated. The crude product was chromatographed on a column (100 cm × 5.5 cm i.d.) of silica gel with petroleum ether as eluent. The compounds 5,6-dichlorodecadiene-1,9, 5,6,9,10-tetrachlorodecene-1 and 1,2,5,6,9,10-hexachlorodecane were eluted successively. If necessary, the chromatography was repeated to reach a purity of more than 95%.

Synthesis of the compounds 9,11,11,11-tetrachloroundecadienes-1,5 and 1,1,1,3,10,12,12,12-octachlorodecenes-6: 15 g 1,5,9-decatriene were refluxed in 150 mL CCl_4 for 6 days, adding 0.5 g AIBN every day during refluxing. After that, the reaction solution was treated with 100 mL of conc. sulfuric acid. The organic phase was separated, washed with water, and rotary evaporated. The residue was dissolved in 50 mL petroleum ether and fractionated over a short silica gel column (50 cm × 5 cm) to separate 9,11,11,11-tetrachloroundecadienes-1,5 from 1,1,1,3,10,12,12,12-octachlorodecenes-6. Thereafter, both compound mixtures were chromatographed with petroleum ether again on silica gel column (100 cm × 2.6 cm), whereas for the separation of 1,1,1,3,10,12,12,12-octachlorodecenes-6, silica gel was activated before use at 150°C for 24 h.

1,2,5,6,9,10-Hexachlorodecane (Fig. 1) [51]

1,2,5,6,9,10-Hexachlorodecane was synthesized by bubbling chlorine gas, at room temperature, into neat 1,5,9-decatriene (Aldrich Chemical Co.) contained in a flask wrapped in aluminum foil to exclude light. The desired compound was the major product of the reaction, as verified by analysis of a 10% solution of the product mixture in hexane by HRGC/MS. In an attempt to prepare an analytical standard, we treated the reaction products as follows. To remove unreacted alkenes, the product mixture (0.25 mL) was mixed with 2 mL of H_2SO_4/HNO_3 (1:1) at 70°C for 20 min and then cooled in an ice bath. After distilled water (5 mL) had been added, the mixture was extracted with hexane (3 × 3 mL). The combined extracts were concentrated to 1 mL and chromatographed on a Florisil column, as described below. Analysis by GC/MS (positive ion TIC) showed the major products to be $C_{10}H_{16}Cl_6$, $C_{10}H_{15}Cl_7$, $C_{10}H_{14}Cl_8$, and $C_{10}H_{13}Cl_9$. Additional peaks in the chromatogram accounted for 7 ± 2% of the TIC.

1,5,5,6,6,10-Hexachlorodecane, A (Fig. 47a) [73]

1,1,1,5-Tetrachloropentane (500 g) and 794 mL of a solution of ammonia in ethanol containing 50 g of ammonia were added to previously reduced platinum oxide

(2.38 g of platinum oxide in 320 mL of alcohol and 10 mL of glacial acetic acid). Reduction was carried out at atmospheric pressure; it proceeded with evolution of heat. In the course of 9 h, 35 L of hydrogen was absorbed. The reaction mixture was diluted with water until the ammonium chloride was dissolved, and it was then filtered. The precipitate of 1,5,5,6,6,10-hexachlorodecane was dissolved in chloroform, and the solution was dried over calcium chloride. The solvent was removed, and low-boiling substances (180 g) were removed under reduced pressure, The hexachlorodecane, which remained behind, was recrystallized from alcohol. The product was 202 g (48.5%) of hexachlorodecane, m.p. 84–85°C.

1,5,5,6,6,10-Hexachlorodecane, B (Fig. 47d) [63]

1) To a solution of 100 g tetrachloropentane in 50 mL ether was added a solution of ethylmagnesium bromide (from 68 g ethyl bromide and 15 g magnesium) m 125 mL ether at such a rate that the ether boiled moderately. Gas consisting of ethane and ethylene was evolved. The mixture was heated and boiled for 15 mm and treated as in the previous experiment (see 5.1.25). Vacuum distillation gave 59.5 g of a fraction with b.p. 78–93°C (9 mm) and 21.4g residue.
Column fractionation gave 44.3 g 1,1,5-trichloropentane with b.p. 82–84°C (8 mm) and 12.4g of the original tetrachloropentane. From the residue was isolated 15.1 g 1,5,5,6,6,10-hexachlorodecane with m.p. 83–84°C and 3.8 g of 1,5,6,10-tetrachlorodecene-5 with b.p. 150–152°C (2.5 mm).
2) To a solution of 40 g 1,1,1,5-tetrachloropentane m 30 mL ether and 2 g anhydrous cobalt chloride was added with mixing ethylmagnesium bromide (from 26 g ethyl bromide and 6 g magnesium) m 70 mL ether. After treatment of the mixture as in the previous experiment, 8.4 g of 1,5,5,6,6,10-hexachlorodecane, 2.3 g of 1,5,6,10-tetrachlorodecene, 12.4 g 1,1,5-trichloropentane, and 5 g of the original tetrachloropentane were obtained.

1,1,1,3,8,10,10,10-Octachlorodecane, 1,1,1,3,9,11,11,11-Octachloroundecane, 1,1,1,3,10,12,12,12-Octachlorododecane (Fig. 5) [6]

The synthesis of 1,1,1,3,10,12,12,12-octachlorododecane and 9,11,11,11-tetrachloroundecene-1 (from 1,9-decadiene), 1,1,1,3,8,10,10,10-octachlorodecane (from 1,7-octadiene), and 8,10,10,10-tetrachlorodecene-1 and 1,1,1,3,9,11,11,11-octachloroundecane (from 1,8-nonadiene) was analogous with the above-described synthesis of 1,2,5,6,9,10-hexachlorodecane (see 5.1.23). Separation of diastereomers could not be achieved under the experimental conditions used. The products were purified on a silica gel column (100 cm × 2.6 cm) with petroleum ether.

1,1-Dichloroundecane, 1,1,5-Trichloroundecanes and 1,1,6-Trichloroundecanes (Figs. 42 and 43) [64]

The reaction product from 1,1,1-trichloroundecane (9.5 g, 40 mmole) was fractionated in vacuum. Fractions obtained were: 1) b.p. 68–70°C (1.5 mm) (1.7 g); 2) b.p. 80–85°C (1 mm) (3.1 g); 3) b.p. 85–89°C (1 mm) (4 g).

From fraction 1, consisting of 95% 1,1-dichloroundecane and 5% 1,1,1-trichloroundecane according to GLC data, a repeat distillation gave 1,1-dichloroundecane b.p. 98°C (1.5 mm). ^{13}C NMR spectrum (δ, ppm): 73.2 (C-1), 43.7 (C-2), 30.1 (C-3), 26.3 (C-4), 29.0 (C-5), 29.3 (C-6), 29.5 (C-7), 26.0 (C-8), 31.9 (C-9), 22.6 (C-10), 13.9 (C-11).

According to GLC, fraction 3 contained 80% 1,1,5- and 1,1,6-trichloroundecane and 20% other isomers.

1,7,7,8,8,14-Hexachlorotetradecane (Fig. 47a) [73]

1,1,1,7-Tetrachloroheptane (83 g) and a solution of 8 g of ammonia m 80 mL of methanol were added to previously reduced 5% Pd/BaSO$_4$ (2 g) in 25 mL of methanol and 1 mL of glacial acetic acid. During the hydrogenation the reaction mixture became warm. After 3 h, when 6.1 L of hydrogen had been absorbed, the absorption of hydrogen ceased. The methanolic solution was filtered from the catalyst and ammonium chloride, and the precipitate was washed with chloroform on the filter. Water was added to the filtrate. The chloroform layer was dried over calcium chloride. After removal of chloroform under reduced pressure the low-boiling products were distilled off (28 g). Recrystallization of the residue from ethanol yielded 32.1 g (45.5%) of 1,7,7,8,8,14-hexachlorotetradecane, m.p. 57–58°C.

1,14-Dichlorotetradecane (Fig. 47a) [73]

Tetrachloroheptane (90 g) was hydrogenated in the presence of 1.5 g of Pd/BaSO$_4$ and 8.1 g of ammonia in 100 mL of methanol for 2 h 30 min, in the course of which 7.5 L of hydrogen was absorbed. The Catalyst and ammonium chloride were filtered off, methanol was distilled off, and low-boiling fractions were removed under reduced pressure. The residue (42 g) was hydrogenated in the presence of 3 g of Pd/BaSO$_4$ and 32 g of diethylamine in 80 mL of ethanol for ten hours. The product was 14.5 g (28%) of 1,14-dichlorotetradecane, b.p. 144–146°C (1.8 mm), together with about 8 g of a mixture boiling at 180–180°C (1 mm) and consisting mainly of tetrachlorotetradecene.

1,1,14,14-Tetrachlorotetradecane (Fig. 47d) [76]

With stirring, to 1.0 g of Mg, activated with iodine, was added a mixture of 9.9 g of 1,1-dichloro-7-bromoheptane and 0.9 g of 1,2-dibromoethane in 80 mL of absolute ether. The reaction started immediately. After heating the mixture for 3 h, followed by the usual workup and vacuum-distillation, we obtained 0.9 g of 1,1-dichloroheptane, 2.9 g of the starting bromide, and 2.7 g of 1,1,14,14-tetrachlorotetradecane with b.p. 171–174°C (<1 mm). The latter after redistillation had b.p. 156–158°C (0.5 mm).

1,9,9,10,10,18-Hexachlorooctadecane (Fig. 47a) [72]

A solution of 6.8 g (0.4 mole) of ammonia in 110 mL of methanol and 106 g (0.4 mole) of 1,1,1,9-tetrachlorononane were added to previously reduced palladium oxide (6 g of 5% $Pd/BaSO_4$ in 50 mL of methanol containing 2 mL of glacial acetic acid. The hydrogenation was carried out in a glass hydrogenation flask at atmospheric pressure. After ten hours the absorption of hydrogen stopped. In all, 4.5 L of hydrogen was absorbed. The catalyst was filtered off and washed with water and chloroform; the chloroform layer was dried over calcium chloride. After distillation of the chloroform we obtained 47.5 g (51.7%) of 1,9,9,10,10,18-hexachlorooctadecane, m. p. 55.5–56°C (from a mixture of alcohol and acetone). In addition we isolated 10 g of nonyl chloride; b.p. 55–56°C (1 mm).

Typical Reaction Conditions for the Ru-Catalyzed Addition of Haloalkanes to oct-1-ene (Fig. 32) [46]

Oct-1-ene (3.4 g, 0.03 mol), haloalkane (0.06 mol) and $[RuCl_2(PPh_3)_3]$ (0.01 g, 1×10^{-5} mol) were loaded into a glass tube with a restriction in the neck to facilitate sealing. The reaction mixture was degassed three times by the freeze-pump-thaw method and the tube was then sealed under vacuum. The tube was then heated to the desired temperature in an oven for times varying between 1.5 and 20 h, after which time it was opened and the contents analyzed by GLC.

References

1. Bayen S, Obbard JP, Thomas GO (2006) Chlorinated paraffins: A review of analysis and environmental occurrence. Environ Int 32:915–929
2. Tomy GT, Stern GA, Muir DCG et al (1997) Quantifying C_{10}–C_{13} Polychloroalkanes in Environmental Samples by High-Resolution Gas Chromatography/Electron Capture Negative Ion High-Resolution Mass Spectrometry. Anal Chem 69:2762–2771
3. El-Morsi TM, Emara MM, Abd El Bary HMH et al (2002) Homogeneous degradation of 1, 2, 9, 10-tetrachlorodecane in aqueous solutions using hydrogen peroxide, iron and UV light. Chemosphere 47:343–348
4. Fisk AT, Cymbalisty CD, Tomy GT et al (1998) Dietary accumulation and depuration of individual C_{10}-, C_{11}- and C_{14}-polychlorinated alkanes by juvenile rainbow trout (Oncorhynchus mykiss). Aquat Toxicol 43:209–221

5. Coelhan M, Saraci M, Lahaniatis ES et al (1998) Contribution to the quantification of C_{10}-chloroparaffines: Part 1. First time quantification of C_{10}-chloroparaffines with purely synthesized chloroalkanes as standards. Fresenius Environ Bull 7:353–60

6. Coelhan M (2003) Synthesis of several single C_{10}, C_{11}, and C_{12} chloroalkanes. Fresenius Environ Bull 12:442–9

7. Tomy GT, Billeck B, Stern GA (2000) Synthesis, isolation and purification of C10–C13 polychloro-n-alkanes for use as standards in environmental analysis. Chemosphere 40:679–83

8. Beaume FN (2005) Bestimmung von C_{10}-Chlorparaffinen mit einem synthetisierten Standard in Lebensmitteln. Ph.D. Thesis, Technical University of Munich, Germany

9. Beaume F, Coelhan M, Parlar H (2006) Determination of C10-chloroalkane residues in fish matrices by short column gas chromatography/electron capture negative ion low resolution mass spectrometry applying single pure and representative synthesized chlorodecanes as standards. Anal Chim Acta 565:89–96

10. Frenzen G, Sippel H, Coelhan M (1999) The relative configuration of a stereoisomer of 1, 2, 5, 6, 9, 10-hexachlorodecane. Acta Cryst. doi:10.1107/S0108270199099965

11. Bart JCJ, Bassi IW, Calcaterra M (1979) Molecular and crystal structure of meso-(RS)-1, 1, 1, 3, 6, 8, 8, 8-octachlorooctane. Acta Cryst B35:2646–2650

12. Bassi IW, Scordamaglia R (1974) The addition halogenation of poly(1-alkenylene)s, 5. X-ray crystal structures of (RS)-1, 2, 5, 6-tetrachlorohexane and (RRSS)-2, 3, 6, 7-tetrachloroctane. Makromol Chem 175:1641–1650

13. Ransley DL (1968) Long-range effects in the alkylation of benzene with dichloroalkanes. J Org Chem 33(4):517–522

14. Tordeux M, Boumizane K, Wakeelman C (1993) Preparation of *gem*-dichloroalkanes from oximes. J Org Chem 58:1939–1940

15. Ransley DL (1969) Long-range effects in the alkylation of benzene with polyhalooctanes. J Org Chem 34(9):2618–2621

16. Jensen SR, Brown WA, Heath E et al (2007) Characterization of polychlorinated alkane mixtures - a Monte Carlo modeling approach. Biodegradation 18:703–717

17. Ramage MP, Eckert RE (1975) Kinetics of the liquid phase chlorination of *n*-dodecane. Ind Eng Chem Fundam 14(3):214–221

18. Turro NJ, Fehlner JR, Hessler DP et al (1988) Photochlorination of *n*-alkanes adsorbed on pentasil zeolites. J Org Chem 53(16):3731–3735

19. Turro NJ, Han N, Lei X et al (1995) Mechanism of dichlorination of *n*-dodecane and chlorination of 1-chlorododecane adsorbed on ZSM-5 zeolite molecular sieves. A supramolecular structural interpretation. J Am Chem Soc 117:4881–4893

20. Dneprovkii AS, Miltsov SA (1986) Mechanisms of free-radical reactions. XIX. Selectivity of the free-radical chlorination of 1-chloroalkanes by N-chloropiperidine. Zh Org Khim (Russ) 22(2):265–269

21. Dneprovkii AS, Miltsov SA, Arbuzov PV (1988) Mechamisms of free-radical reactions. XXIV. Quantitative description of the polar effects of substituents on the kinetics of the free-radical chlorination of aliphatic compounds by N-chloropiperidine. Zh Org Khim (Russ) 24(10):2026–2037

22. Kharasch MS, Jensen EV, Urri WH (1945) Addition of carbon tetrachloride and chloroform to olefins. Science 102:128

23. Kharasch MS, Urri WH, Jensen EV (1945) Addition of derivatives of chlorinated acetic acids to olefins. J Am Chem Soc 67:1626

24. Kharasch MS, Jensen EV, Urri WH (1945) The addition of phosphorous trichloride to olefins. J Am Chem Soc 67:1863–1864

25. Kharasch MS, Jensen EV, Urri WH (1946) Addition of carbon tetrabromide and bromoform to olefins. J Am Chem Soc 68:154–155

26. Kharasch MS, Jensen EV, Urri WH (1947) Reactions of atoms and free radicals in solution. X. The addition of polyhalomethanes to olefins. J Am Chem Soc 69:1100–1105

27. ETs C, Kuz'mina NA, Freidlina RKh (1970) A new reaction of the CCl_3 group in polychlorohydrocarbons. Izv Akad Nauk SSSR Ser Khim 10:2343–2351
28. ETs C, Kuz'mina NA, Rozhkova MA et al (1982) Reaction of 1, 1, 1, 3-tetrachloropropane with 1-hexene, initiated by Fe(CO)5, Mo(CO)6 and Mn2(CO)10. Izv Akad Nauk SSSR Ser Khim 6:1345–1349
29. Kamyshova AA, ETs C, Freidlina RKh (1973) Reaction of 1, 1, 1-trichloroethane or carbon tetrachloride with allyl chloride, initiated by the Fe(CO)5-isopropanol system. Izv Akad Nauk SSSR Ser Khim 9:2004–2008
30. Kruglova NV, Freidlina RKh (1984) Reduction of 1, 1, 1-trichloroethane and its addition to 1-heptene initiated by tert-butyl peroxide or Fe(CO)5 in the presence of triethylsilane. Izv Akad Nauk SSSR Ser Khim 2:388–392
31. Rybakova NA, ETs C, Freidlina RKh (1979) Addition of 1, 1, 1, 3-tetrachloropropane to 1-hexene, initiated by system Fe(CO)5 + HMPA + p-ClC6H4SO2NCl2. Izv Akad Nauk SSSR Ser Khim 11:2618–2620
32. Zhiryukhina NP, Kamyshova AA, ETs C et al (1983) Synthesis of polychloroalkanes with several different chlorine-containing groups. Izv Akad Nauk SSSR Ser Khim 1:152–157
33. Freidlina RKh, Osipov BN (1971) Telomerization of ethylene with 1, 1, 1, 3-tetrachloropropane in the presence of iron pentacarbonyl and isopropyl alcohol. Izv Akad Nauk SSSR Ser Khim 12:2837–2839
34. Freidlina RKh, Kuz'mina NA, Kamyshova AA et al (1977) Reaction of *gem*-trichloroalkanes with unsaturated compounds in presence of $Fe(CO)_5$ and a cocatalyst. Izv Akad Nauk SSSR Ser Khim 1:174–177
35. Kuz'mina NA, ETs C, Freidlina RKh (1978) Telomerization of 2-chloropropene with 1, 1, 1-trichloroethane. Izv Akad Nauk SSSR Ser Khim 12:2827–2828
36. ETs C, Freidlina RKh (1972) Telomerization of butadiene with 1, 1, 1-trichloroethane and 1, 1, 1, 3-tetrachloropropane. Izv Akad Nauk SSSR Ser Khim 2:468–470
37. Destarac M, Bessiere J-M, Boutevin B (1998) Atom transfer radical polymerization of styrene initiated by polychloroalkanes and catalyzed by CuCl/2, 2-bipyridine: A kinetic and mechanistic study. J Polymer Sci A: Polym Chem 36:2933–2947
38. Burton DJ, Kehoe LJ (1970) Copper chloride-ethanolamine catalyzed addition of polyhaloalkanes to 1-octene. J Org Chem 35:1339–1342
39. Freidlina RKh, ETs C (1962) Use of oxidation-reduction systems for the initiation of 1-octene with carbon tetrachloride. Izv Akad Nauk SSSR Otd Khim Nauk 10:1788–I788
40. Davis R, Groves IF (1982) The reaction of tetrachloromethane with oct-1-ene in the presence of $[Mo_2(CO)_6(\eta\text{-}C_5H_5)_2]$ and other transition-metal complexes. J Chem Soc Dalton Trans 2281-2287
41. Davis R, Durrant JLA, Khazal NMS et al (1990) The addition of halogenocarbons to alkenes in the presence of $[Fe_2(CO)_4(\eta\text{-}C_5H_5)_2]$ and related complexes. J Organomet Chem 386:229–239
42. Shvo Y, Green R (2003) Addition of α-polyhalides to olefins under mild reaction conditions, catalyzed by Mo(CO)6. J Organomet Chem 675:77–83
43. Davis R, Khazaal NMS, Maistry V (1986) Addition of halogenocarbons or dihydrogen to alkenes in the presence of $[Fe_2(CO)_4(\eta\text{-}C_5H_5)_2]$: Catalysis by a dinuclear species. J Chem Soc Chem Commun 1387–1389
44. Davis R, Furze JD, Cole-Hamilton DJ et al (1992) The mechanism of the addition of haloalkanes to alkenes in the presence of $[RuH_3(SiMe_2Ph)(PPh_3)_3]$ and $[RuH_2(PPh_3)_4]$. J Organomet Chem 440:191–196
45. Matsumoto H, Nakano T, Nagai Y et al (1978) Catalysis by phosphine complexes of silylruthenium hydrides of the addition of carbon tetrachloride to 1-octene. Bull Chem Soc Jap 51(8):2445–2446
46. Bland WJ, Davis R, Durrant JLA (1985) The mechanism of the addition of haloalkanes to alkenes in the presence of dichlorotris(triphenylphosphine)-ruthenium(II) [RuCl2(PPh3)3]. J Organomet Chem 280:397–406

47. Richel A, Demonceau A, Noels AF (2006) Electrochemistry as a correlation tool with the catalytic activities in [RuCl$_2$(p-cymene)(PAr$_3$)]-catalyzed Kharasch additions. Tetr Lett 47:2077–2081

48. Tutusaus O, Delfosse S, Demonceau A et al (2003) Kharasch addition catalysed by half-sandwich ruthenium complexes. Enhanced activity of ruthenacarboranes. Tetr Lett 44:8421–8425

49. Freidlina RKh, Vasil'eva EI (1958) Hydrolysis of polyhalohydrocarbons containing the CHal$_3$ or CCl$_2$=CH group. Russ Chem Bull 7(1):32–35

50. Saotome K, Komoto H, Yamazaki T (1966) The synthesis of α, ω-disubstituted higher alkanes from α, α, α, ω-tetrachloroalkanes. Bull Chem Soc Jap 39:480–484

51. Mayer Z (1974) Thermal decomposition of polyvinyl chloride and of its low-molecular-weight model compounds. Polymer Rev 10:263–292

52. Smirnov VV, Tarkhanova IG, Kokorin AI et al (2005) Catalytic activity of immobilized transition metal complexes with monoethanolamine in carbon tetrachloride addition to multiple bonds. Kinet Katal 46(6):909–914

53. Tarkhanova IG, Smirnov VV, Rostovshchikova TN (2001) Distinctive characteristics of carbon tetrachloride addition to olefins in the presence of copper complexes with donor ligands. Kinet Katal 42(2):216–222

54. Tarkhanova IG, Gantman MG, Chizhov AO et al (2006) Addition of tetrachloromethane to oct-1-ene initiated by amino alcohols. Russ Chem Bull Int Ed 55(9):1624–1630

55. Mitani M, Kiriyama T, Kuratate T (1994) Addition reaction of polychloro compounds to carbon-carbon multiple bonds catalyzed by semiconductor particles under photoirradiation. J Org Chem 59:1279–1282

56. Joyce RM, Hanford WE, Harmon J (1948) Free radical-initiated reaction of ethylene with carbon tetrachloride. J Am Chem Soc 70(5):2259–2262

57. Asscher M, Levy E, Rosin H et al (1963) Telomerization of ethylene and carbon tetrachloride. Novel initiating system. Ind Eng Chem Prod Res Dev 2(2):121–126

58. Nesmeianov AN, Kareptian ShA, Freidlina RKh (1956) Synthesis of higher α, α, α, ω-tetrachloroalkanes and 1, 1, 1-trichloroalkanes. Proc Acad Sci USSR 109(4):791

59. Karapet'yan ShA, Zelenskaya LG, Kruglova NV et al (1966) Infrared spectra, molecular volumes and refractions of α, α, α, ω-tetrachloroalkanes and α, α, ω-trichloroalkenes, containing five to eleven carbon atoms. Izv Akad Nauk SSSR Ser Khim 8:1355–1360

60. Nesmeyanov AN, Semenov NA (1959) Preparation of α, α, ω-trichloroalkenes from α, α, α, ω-tetrachloroalkanes. Russ Chem Bull 8(12):2022–2024

61. Freidlina RKh, Belyavskii AB (1961) Telomerization of ethylene and carbon tetrachloride or chloroform in the presence of the hexacarbonyls of chromium, molybdenum and tungsten. Izv Akad Nauk SSSR Otd Khim Nauk 1:177–178

62. Kuz'mina NA, ETs C, Freidlina RKh (1977) Telomerization of vinylidenechloride with 1, 1, 1, 3-tetrachloropropane. Izv Akad Nauk SSSR Ser Khim 4:921–924

63. Zakharkin LI (1963) Interaction of ethylmagnesium bromide with 1, 1, 1, 5-tetrachloropentane. Izv Akad Nauk SSSR Otd Khim Nauk 5:939–941

64. Kiseleva LN, Rybakova NA, Kuz'mina NA et al (1981) Hydrogen migration in 1, 1-dichloroalkyl radicals. Izv Akad Nauk SSSR Ser Khim 9:2095–2098

65. Kuz'mina NA, Zhiryukhina NP, ETs C et al (1981) Reduction of α, α, α-trichloromethylcompounds in the presence of metal carbonyls. Izv Akad Nauk SSSR Ser Khim 9:2090–2094

66. Petrova RG, Freidlina RKh (1970) The reduction of α, α, α, ω-tetrachloroalkanes by the action of thiols in the presence of iron compounds. Izv Akad Nauk SSSR Ser Khim 7:1574–1577

67. Rybakova NA, Kiseleva LN (1981) Radical transformations of 1, 1, 1, 3-tetrachlorononane caused by initiating systems based on Fe(CO)5 and containing various hydrogen donors. Izv Akad Nauk SSSR Ser Khim 7:1636–1638

68. Kost TA, Freidlina RKh (1967) Catalytic hydrogenation of polychloroalkenes containing a CCl$_3$C-CH$_2$ or CCl$_2$=CCHCl group. Izv Akad Nauk SSSR Ser Khim 12:2715–2719

69. Kiseleva LN, Rybakova NA, Freidlina RKh (1983) Reactions of 1, 3, 3, 5-Tetrachloropentane, initiated by Fe, Mn, and Mo carbonyl in the presence of dimethylformamide. Izv Akad Nauk SSSR Ser Khim 10:2408–2410

70. Kiseleva LN, Rybakova NA, Freidlina RKh (1986) Hydrogenolysis of the C-C1 bond of the dichloromethyl group in polychloroalkanes, initiated by $Fe(CO)_5$, $Mn_2(CO)_{10}$, or *tert*-butyl peroxide in combination with triethylsilane. Izv Akad Nauk SSSR Ser Khim 5:1136–1138

71. Rybakova NA, Kiseleva LN, Freidlina RKh (1984) Chemical transformations of 1, 3, 3, 5-tetrachlorononane initiated by Fe(CO)5, Me(CO)6 and Mn2(CO)10 systems. Izv Akad Nauk SSSR Ser Khim 11:2623–2626

72. Freidlina RKh, Kost TA (1960) Action of nucleophilic reagents on compounds of the type [Cl $(CH_2)nCCl=]_2$. Izv Akad Nauk SSSR Otd Khim Nauk 8:1887–1390

73. Nesmeyanov AN, Zakharkin LI, Kost TA (1955) Synthesis of dodecanoic and hexadecanoic acids. Izv Akad Nauk SSSR Otd Khim Nauk 4:585–590

74. Mubarak MS, Peters DJ (1995) Electrochemical reduction of 1, 6-dihalohexanes at carbon cathodes in dimethylformamide. J Org Chem 60:681–685

75. Freidlina RKh, Kost TA (1957) Action of nucleophilic reagents on compounds of the type [Cl $(CH_2)nCCl_2-]_2$. Russ Chem Bull 6(5):656–658

76. Kruglova NV, Freidlina RKh (1972) Reaction of α, α-dichloro-ω-bromoalkanes and α, α-dichloro-ω-bromoalkenes with benzylmagnesium and allylmagnesium halides. Izv Akad Nauk SSSR Ser Khim 4:886–890

Analysis of Chlorinated Paraffins in Environmental Matrices: The Ultimate Challenge for the Analytical Chemist

Gregg T. Tomy

Abstract Commercial chlorinated paraffins (CPs) are derived from the free radical chlorination of n-alkane mixtures. Starting mixtures used in the synthesis fall into three categories: C_{10}–C_{13} (short); C_{14}–C_{17} (medium) and C_{20}–C_{30} (long). This results in complex mixtures containing significant numbers of constitutional and optical isomers. It is this complexity that makes analysis of CPs extremely challenging. Modern analytical methods employ either single or multi-dimensional gas chromatography coupled to mass spectrometric detectors operated in the negative ion mode. This chapter discusses the advances that have been made in the analysis of CPs in environmental samples with a focus on modern analytical techniques.

Keywords Chlorinated paraffins, Short chain chlorinated paraffins (SCCPs), Medium chain chlorinated paraffins (MCCPs), Analytical methods, Mass spectrometry

Contents

G.T. Tomy
Research Scientist, Arctic Aquatic Research Division, Fisheries and Oceans Canada, 501 University Crescent, Winnipeg R3T 2N6, MB, Canada
e-mail: gregg.tomy@dfo-mpo.gc.ca

J. de Boer (ed.), *Chlorinated Paraffins*, Hdb Env Chem (2010) 10: 83–106,
DOI 10.1007/698_2009_39, © Springer-Verlag Berlin Heidelberg 2010,
Published online: 12 January 2010

1 Introduction

Chlorinated paraffins (CPs) were first produced in the 1930s for medicinal purposes. Scheer (1944) reported that a commercial antiseptic solution in the form of a chlorocosane, i.e., a solution containing 20 carbon atoms, was used during World War I [1]. Later, during World War II, CPs were used as flame retardants and were applied to tent canvases and other textile materials [2].

Today, commerical CP formulations are synthesized by the crude chlorination of n-alkane feedstocks with molecular chlorine under forcing conditions, e.g., high temperatures and/or UV irradiation. The extent and conditions of the chlorination employed ultimately depend on the desired application [1, 3]. Because the n-alkane feedstocks are derived from petroleum fractions, the end product is a mixture of carbon chain lengths and chlorination. Commercial CP mixtures fall into three categories: C_{10}–C_{13} (short), C_{14}–C_{17} (medium), and C_{20}–C_{30} (long). The mixtures are further subcategorized on the basis of their weight content of chlorine: 40–50%, 50–60%, and 60–70%.

Owing to the varying carbon chain length and chlorine percentages of technical mixtures, CPs provide a range of properties for different applications. In general, CPs are used where the demand for chemical stability is high [4]; common applications include high temperature lubricants, plasticizers, and flame retardants, and as additives in adhesives, paints, rubber, and sealants [5, 6].

This chapter will attempt to cover the analytical methods used for measuring CPs in environmental samples. A recent survey of the literature has revealed that four comprehensive reviews on analytical methodologies for CP analysis were written and published in the last 3 years [7–10]. Some overlap between this work and the contents of the review papers is therefore unavoidable.

1.1 Complexity of Industrial Mixtures

Free radical halogenation of alkanes under forcing conditions is typically reactive, and the substitution of a hydrogen atom by a chlorine atom is generally not site-specific. Furthermore, the n-alkanes used as starting materials in the industrial process generally consist of a mixture of homologues. The resulting synthetic mixture contains a range of carbon chain length and varying degrees of chlorination. It is this inherent complexity that makes the analysis of CPs in environmental samples extremely challenging.

To illustrate the complexity numerically, Tomy et al. (1997) and Shojania (1999) derived mathematical equations to calculate the theoretical number of possible constitutional isomers of CPs [11, 12]. For a CP of the general formula, $C_nH_{2n+2-z}Cl_z$, and assuming no more than one chlorine atom on any carbon, the number of constitutional isomers is given by:

Table 1 The number of constitutional isomers calculated for $C_nH_{2n+2-z}Cl_z$ by assuming no more than one bound Cl atom on any C atom

$n\rightarrow$ $z\downarrow$	10	11	12	13
1	5	6	6	7
2	25	30	36	42
3	60	85	110	146
4	110	170	255	365
5	126	236	396	651
6	110	236	472	868
7	60	170	396	868
8	25	85	255	651
9	5	30	110	365
10	1	6	36	146
11		1	6	42
12			1	7
13				1

$$N = \frac{1}{2}[\{n!/z!(n-z)!\} + s] \tag{1}$$

where s = the number of symmetrical isomers. Four different cases arise:

1. n even, z even: $s = \{\frac{1}{2}\,n\}!/\{\frac{1}{2}\,z\}!\{\frac{1}{2}\,n - \frac{1}{2}\,z\}!$;
2. n even, z odd: $s = 0$;
3. n odd, z even: $s = \{\frac{1}{2}\,(n-1)\}!/\{\frac{1}{2}\,(n-1) - \frac{1}{2}\,z\}!$;
4. n odd, z odd: $s = \{\frac{1}{2}\,(n-1)\}!/\{\frac{1}{2}\,(z-1)\}!\,\{\frac{1}{2}\,(n-1) - \frac{1}{2}\,(z-1)\}!$

One caveat of the equations is that there can be no more than one chlorine atom bound to any carbon atom. This was imposed because, although free-radical chlorination has low positional selectivity, a second chlorine atom does not readily substitute for a hydrogen at a carbon already bound to chlorine [13–15]. Table 1 shows the number of constitutional isomers for a number of chlorinated alkanes.

For a technical short chain chlorinated paraffin (SCCP) mixture containing 60% chlorine by weight, the theoretical number of congeners (defined as constitutional isomers and homologues) is 4,200 [11, 16]. It should be noted that the complexity would actually be an order of magnitude greater than that indicated in Table 1 because chlorine substitution at a secondary carbon atom usually produces a chiral carbon atom so that enantiomers and diastereoisomers would be generated.

The complexity can be further illustrated by the appearance of a chromatogram of technical formulations derived using a capillary gas chromatograph column. Figure 1 shows the total ion chromatogram of technical SCCP (top panel) and medium chain CP (MCCP, bottom panel) mixtures containing 60 and 53% of chlorine by weight, respectively, obtained using high resolution gas chromatography (GC).

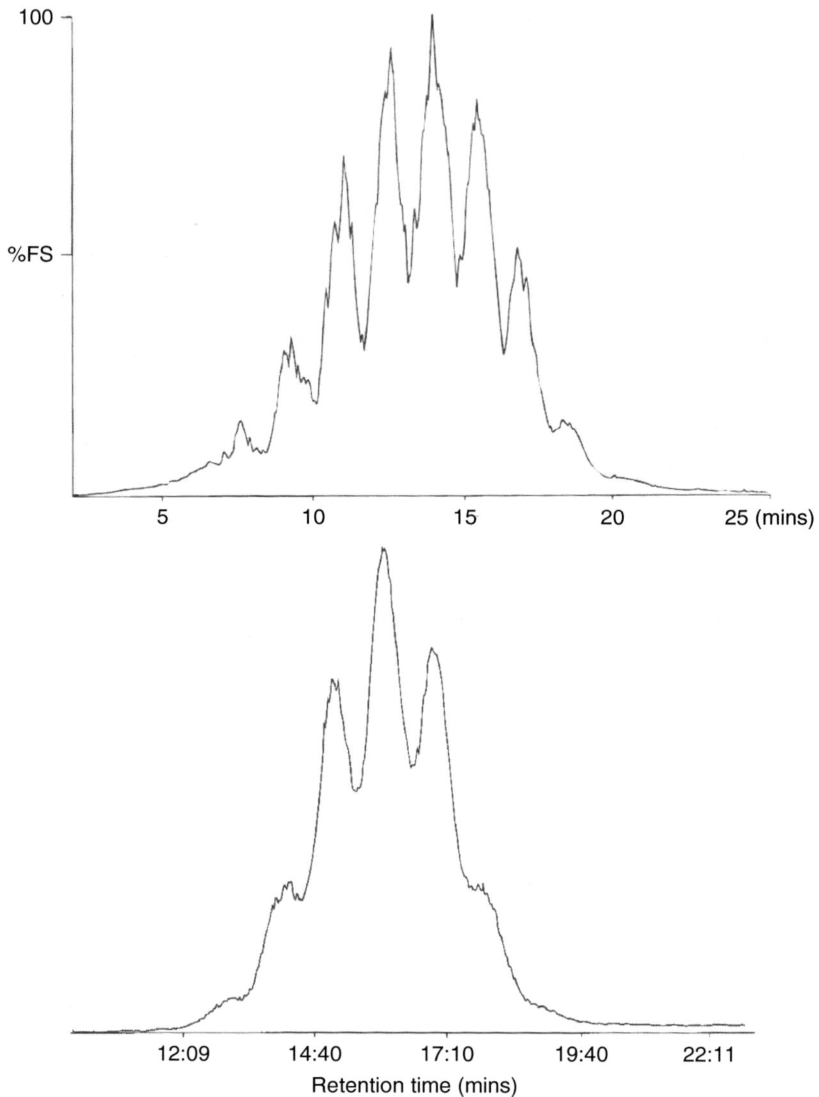

Fig. 1 Total ion chromatograms of an SCCP (*top panel*) and MCCP (*bottom panel*) technical mixture containing 60 and 53% Cl by weight, respectively. Separation obtained using a DB-5MS (dimension: 30 m × 0.25 μm; 0.25 μm stationary film thickness) capillary gas chromatography (GC) column [11, 16]

Based on the large number of congeners present in CP mixtures and in environmental samples, it is not too surprising that the analysis of CPs can be quite challenging. In fact, Coquery et al. (2005) described the short-chain chlorinated paraffins (SCCPs) *as the most challenging group of substances to analyze and quantify* [17].

Table 2 Number of congeners for a suite of environmentally relevant contaminants

Name	# Congeners
Polychlorinated biphenyls	209
Polychlorinated naphthalenes	75
Polychlorinated bornanes	32,768[a]
Polychlorinated dibenzo-p-dioxins	75
Polychlorinated dibenzofurans	135
Polybrominated diphenyl ethers	209

[a]In reality, the actual # of compounds found in the technical mixture is closer to 1,000 (excluding enantiomers) [71]

To put the number of congeners of CPs into perspective, the numbers of congeners of other environmentally relevant contaminants are shown in Table 2.

2 Sample Extraction and Clean-Up

In general, the extraction and clean-up of CPs in environmental matrices is similar to those of other lipophilic organo-*chlorine* and -*bromine* compounds. Furthermore, precautions taken in trace laboratory analysis of other organohalogented compounds like cleanliness of glassware, use of high purity solvents, heat treating adsorbent materials like silica gel and Florisil, and eliminating the use of plastic, also apply to CPs.

Apolar solvents like dichloromethane (DCM), hexane (Hex), and mixtures of DCM:Hex and of ethyl acetate/cyclohexane, Hex/acetone have all been used successful to extract CPs from biota and sediment samples [11, 18–27].

Soxhlet extraction remains the most popular means of extracting CPs from solid samples [11, 18, 19, 23, 25, 28]. This classical technique is robust, easy to use, and inexpensive and has been applied successfully to the analysis of other organohalogenated compounds. Drawbacks to the technique are lengthy extraction times (typically greater than 6 h) and the use of large volumes of solvents. Newer techniques like pressurized fluid extraction (PLE, or accelerated solvent extraction (ASE)) and microwave assisted extraction (MAE) have been shown to mitigate these two factors.

Parera et al. (2004) were the first to compare the extraction efficiencies of SCCPs from sediment using MAE and Soxhlet and showed extraction recoveries to be comparable [29]. Using ASE, Tomy et al. (1999) found recoveries of MCCPs in the range of 79–108% in spiked sodium sulfate (a surrogate for solid samples) [16]. Marvin et al. (2003) also used ASE for extraction of SCCPs in sediment; surrogate recoveries of greater than 75% were reported [20]. Nilsson et al. (2001) also applied ASE to the extraction of household waste and found greater than 90% recoveries of MCCPs [30].

Three methods have been used for extracting or pre-concentrating CPs in water (a) solid-phase extraction (SPE) [19, 23, 31, 32], (b) solid-phase microextraction (SPME) [32, 33] and (c) liquid–liquid extraction [34].

Perhaps the biggest challenge in the analysis of CPs is removal of co-extracted compounds many of which can interfere with the quantification of CPs themselves. Gel permeation chromatography [11, 16, 23, 30, 31, 35–38], concentrated sulphuric acid treatment [25] or sulphuric acid-silica gel column chromatography [39] have all been used to remove co-extracted lipids from extracts. Adsorption chromatography on Florisil, silica, and alumina have all been used to further separate CPs from interfering compounds [11, 16, 29, 34–36, 39, 40]. It should be noted that Reiger and Ballschmiter (1995) have cautioned against the use of alumina as they have showed that dehydrochlorination of CPs on alumina during the adsorption process can lead to partial or total destruction of CPs [41]. Using a high energy mercury lamp, Fridén et al. (2005) were able to eliminate halogenated aromatic interfering compounds like DDT, HCB, and higher chlorinated PCB congeners in their extract [38].

3   Instrumental Analysis

3.1   *Chromatographic Separation*

The complexity of CP mixtures precludes the complete separation of individual congeners using either high performance liquid chromatography (HPLC) or GC. Tomy attempted to resolve a technical SCCP mixture by reverse-phase HPLC using a C_{18}-column [42]. Little separation was obtained using this method likely because of the complexity of technical mixtures and the small number of theoretical plates of HPLC columns.

Even with the improved number of theoretical plates relative to HPLC columns, injection of a technical material onto a single non-polar capillary GC column results in a broad hump, eluting over several minutes (see Fig. 1). Throughout the elution period there can be individual broad peaks. It is thought that the congeners present at greater concentrations in the mixture give rise to the individual broad peaks, while the underlying broad hill result from congeners present at smaller concentrations[11, 43, 44].

Various non-polar stationary phases have been used in an attempt to improve on the separation: 5% phenyl- and 100% -methylpolysiloxane have been reported, and even slightly more polar phases like 35% phenyl-methylpolysiloxane have been used [11, 19–21, 30, 45, 46]. Looser and Ballschmiter (1999) were unable to attain any appreciable improvement in resolution on a highly thermally stable stationary phase Optima δ-3 capillary column [47].

While most work has been done on capillary columns, 30-m in length, of 0.25 μm film-thickness and 0.25 mm internal diameter, Coelhan (1999) first

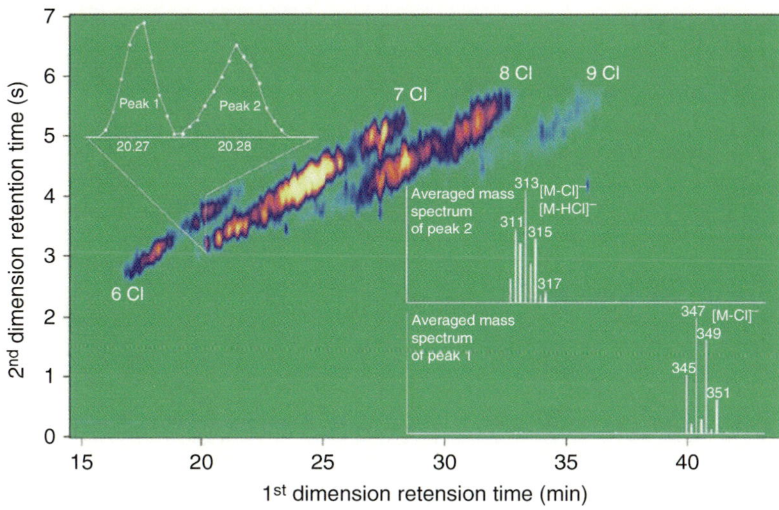

Fig. 2 Full scan (m/z 210–490) GC×GC-ECNI quadrupole MS ion chromatogram of a mixture of polychlorinated decanes with an average Cl content of 65% by weight. Separations performed using a DB-1×007-65HT column combination [52]

proposed the use of shorter columns (65 cm to 15 m) [48]. The rationale is that with very short columns, SCCPs can be eluted as a single peak, thereby reducing analysis times and improving analytical sensitivity [48]. Others have recognized the benefit of this approach and have adopted shorter columns in their analytical procedure [21, 49]. The obvious drawback to using shorter capillary columns is that there is an increased risk of co-elution of CPs with other co-extracted organohalogenated compounds. As such, more thorough sample clean-up procedures are necessary.

In a series of studies, Korytár et al. (2005a–c) applied comprehensive two-dimensional (2D) GC (GC×GC) to improve on CP separations achieved using single-capillary columns [7, 50–52]. The authors first tested their GC×GC analysis on a mixture of polychlorinated decanes (C_{10}) with an average chlorine content of 65% by weight [52]. Figure 2 shows the separation achieved using a DB-1×007-65HT column combination. While the separation of the congeners was not complete, an ordered structure with four parallel groups of peaks corresponding to chlorine content was observed.

A follow-up study by Korytar et al. (2005a) on technical SCCP formulations, using a DB-1×007-65HT combination of stationary phases, showed that separation of CP congeners with the same carbon chain length was based on the number of chlorine atoms (see Fig. 3, [50]). Interestingly, when the mixtures were analyzed, ordered structures comprising compounds having the same number of carbon plus chlorine atoms (e.g., $C_{10}Cl_8$ and $C_{11}Cl_7$) were observed. Furthermore, the authors were able to achieve partial separation of short-, medium-, and long-chain CPs in environmental samples (see Fig. 4, [50]).

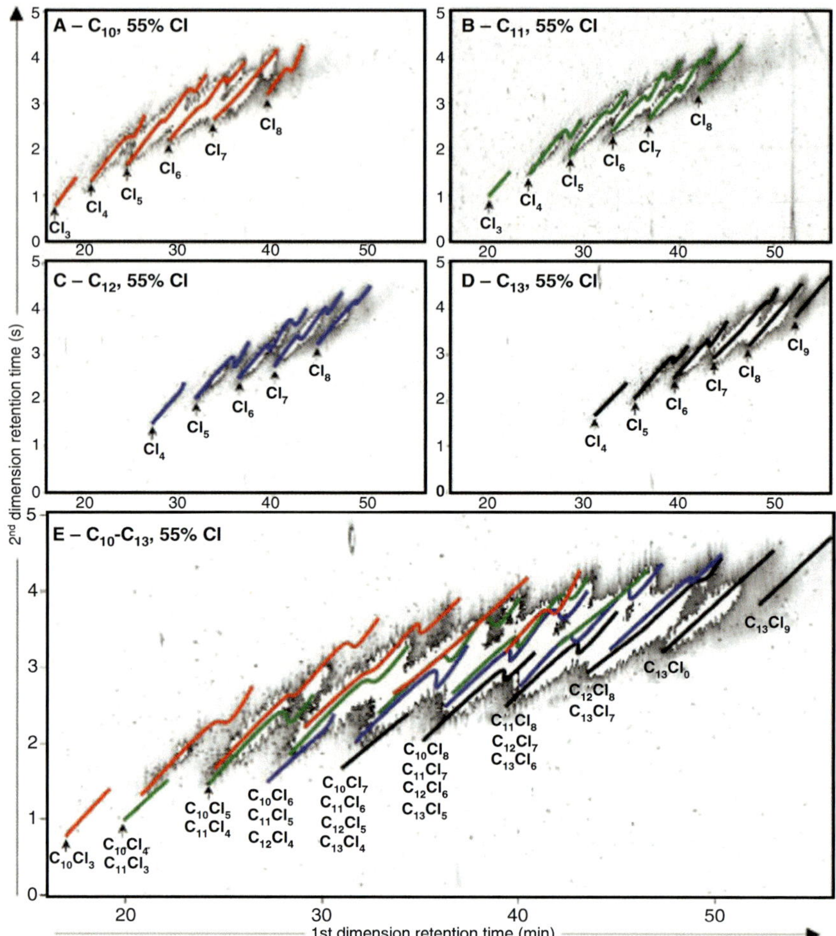

Fig. 3 GC×GC–ECNI-TOF-MS chromatograms of polychlorinated (**a**) decanes, (**b**) undecanes, (**c**) dodecanes, (**d**) tridecanes and (**e**) C$_{10}$–C$_{13}$ technical mixture, all with 55% (w/w) Cl content, obtained on DB-1×007-65HT column combination. *Lines* indicate the positions of apices within the bands [50]

In an extension of their own work, Korytár et al. (2005b) showed that by using a combination of a DB-1 and 65% phenylmethylpolysiloxane capillary columns, it was possible to separate polybrominated diphenyl ethers (PBDEs) and toxaphenes from CPs [51].

The use of 2D-GC is clearly a powerful and promising approach to CP analysis in complex matrices. However, many laboratories have been slow to embrace this type of instrumentation partly because of its specialized nature.

An intriguing approach of carbon skeleton GC has also been applied to the analysis of CPs [44, 53, 54]. In this technique, catalytic reduction of halogenated

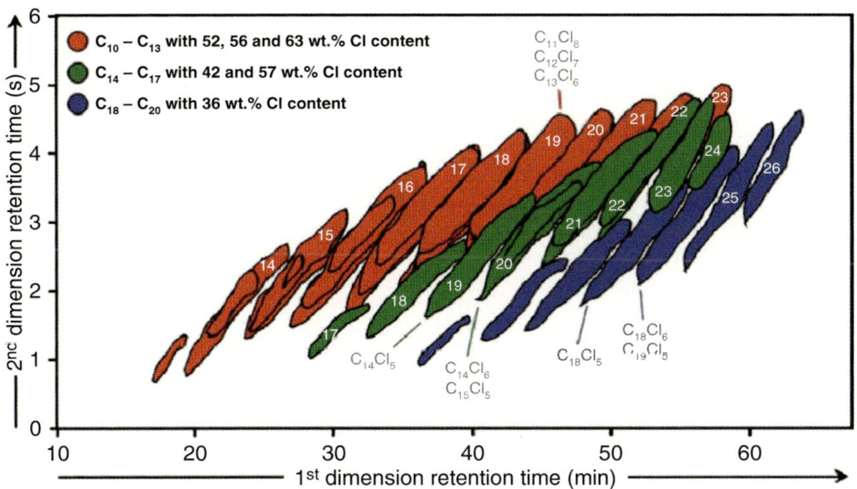

Fig. 4 Overlay of GC×GCECNI-TOF-MS chromatograms of *red* SCCP, *green* MCCP, and *blue* LCCP mixtures with different chlorine content, obtained on DB-1×007-65HT column combination. Different chlorine contents are indicated by additional contours. *White* numbers indicate the number of C + Cl atoms of the compounds present in the bands [50]

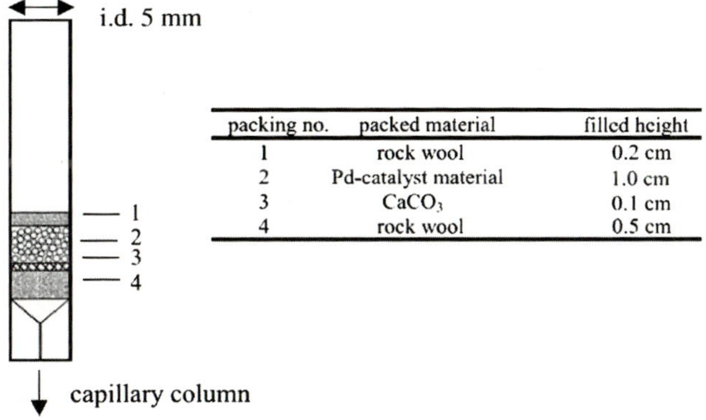

packing no.	packed material	filled height
1	rock wool	0.2 cm
2	Pd-catalyst material	1.0 cm
3	CaCO$_3$	0.1 cm
4	rock wool	0.5 cm

Fig. 5 Typical arrangement of GC injector for carbon skeleton GC [53]

compounds takes place directly inside the injector port which is packed with a palladium catalyst. Figure 5 shows a typical packing arrangement for the GC-injector [53]. For CPs, the reduction reaction results in the formation of corresponding non-chlorinated *n*-alkanes (see Fig. 6). While all information on the degree of chlorination is lost during the reductive process, the procedure does allow for the determination of total amounts of the individual *n*-alkane homologues [44]. Routine detection of the well-resolved *n*-alkanes can be achieved using simple flame ionization detectors.

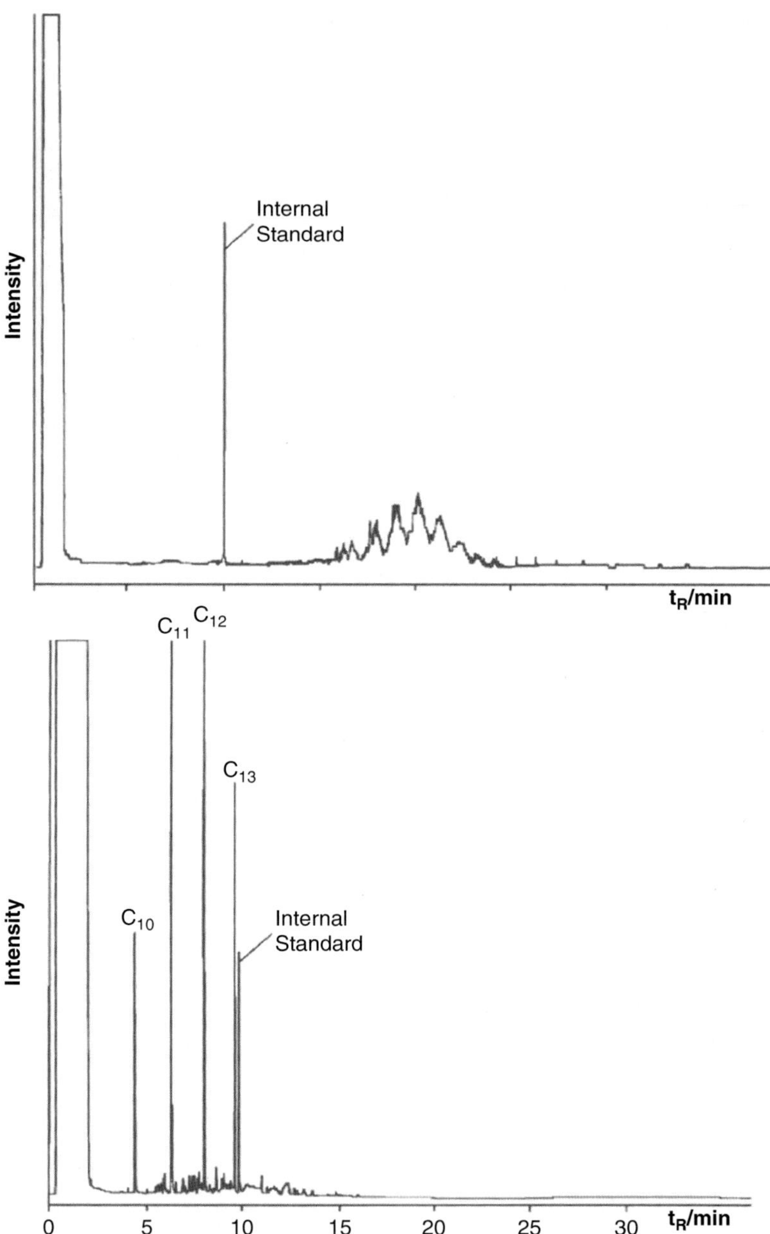

Fig. 6 GC-FID total ion chromatogram of a SCCP containing 56% Cl by weight before (*top panel*) and after (*bottom panel*) catalytical dehydrochlorination using a Pd-catalyst injection port [53]

3.2 Detection

Owing to their high electronegativity, detection of CPs is usually achieved either by electron capture or by mass spectrometric detectors. While electron ionization (EI) has been applied to the analysis of CPs, electron capture negative ionization (ECNI) has been used much more frequently. This section will cover detection methods that have been used in CP analysis.

3.2.1 Electron Capture Detector

With its high sensitivity to electronegative compounds, low cost, and ease of use, it is not surprising that GC coupled to electron capture detectors (ECDs) have been used to detect CPs [30, 33, 38, 51, 55, 56]. However, the elution of CPs over a broad retention time range and the general lack of selectivity of ECDs to other electro-negative compounds require that efficient clean-up procedures be in place to avoid chemical interferences from other co-extractives. Alternatively, by improving on the GC-separation using 2D-GC, Korytár et al. (2005b) were able to employ a micro-ECD as a detector [51]. In general, ECD has not been widely adopted as detector of choice for routine analysis of CPs.

3.2.2 Mass Spectrometric Detectors

The first report on the use of mass spectrometry (MS) for detection of CPs was by Gjøs and Gustavsen in 1982 [57]. Since that time, numerous reports have appeared on the application of MS for detection of CPs. This section will cover the different ionization methods that have been reported on in the literature.

Electron Capture Negative Ionization

The general appearance of an ECNI mass spectrum of a chlorinated alkane is shown in Fig. 7. At low ion source temperatures ca. 100–120°C, mass spectra are typically dominated by the ion cluster corresponding to the $[M-Cl]^-$ ion fragment. With increasing source temperatures, ions corresponding to Cl_2^- (m/z 70) and HCl^- (m/z 71), dominate mass spectra. Perhaps the most important feature of the ECNI mass spectra of CPs is that there is little ion fragmentation. This means that unlike EI where there can typically be many fragment ions, the total ion current imparted to the molecule when first ionized under ECNI conditions is not distributed to many other ions. This leads to enhanced sensitivity under ECNI especially when dominant fragment ion is monitored.

Gjøs and Gustavsen (1982) were the first to report on the use of ECNI-MS on the analysis of CPs [57]. With methane as their moderating gas, the authors found that their ECNI mass spectra were dominated by ions at odd mass with the major fragment

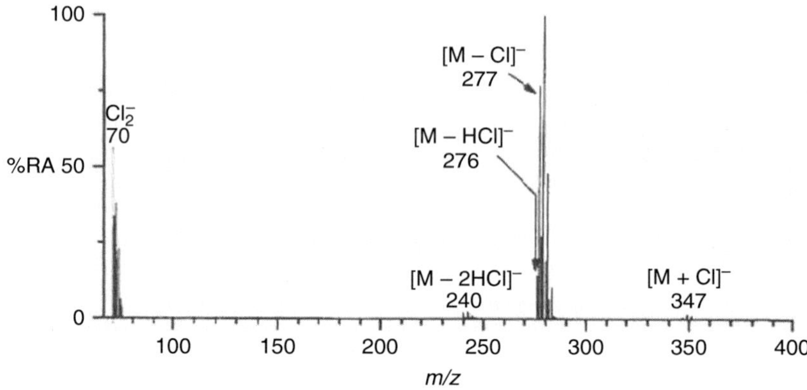

Fig. 7 Electron capture negative ionization (ECNI) mass spectrum of pentachlorodecane [60]

ions assigned as [M–H]⁻· but contributions from [M−Cl]⁻ and [M−HCl−Cl]⁻ were also present. Introduction of samples into the MS was not reported.

Müller and Schmid (1984) built on the earlier work of Gjøs and Gustavsen by introducing CP mixtures into a low resolution ECNI-MS ion source via a GC fitted with a 15 m capillary column [58]. The appearance of the chromatogram is remarkably similar to what is generated today by most laboratories.

Jansson et al. (1991) developed a low resolution ECNI-MS method in the selected ion mode (SIM) [59]. CPs were first selectively removed from other potential interferences by GPC and detection was based on the response of the $Cl_2^{-\cdot}$ (m/z 70) ion. As noted earlier, this ion dominates the ECNI mass spectrum of individual CP congeners at high mass temperature [60].

Metcalfe-Smith et al. (1995) employed a low resolution ECNI-MS method for the analysis of CPs in the full-scan mode [61]. The use of the full-scan method enabled the ion response from CPs to be discriminated against the response from other interfering compounds like PCBs.

The first report on the use of ECNI high resolution mass spectrometry (HRMS) was in 1997 [11]. This method was based on measuring the [M−Cl]⁻ ions of each CP congener in the SIM mode at a resolving power of 12,000. Because of the large number of ions monitored, retention time windows were used where select ions could be monitored over short time periods as opposed to over the entire elution period. Figure 8 shows an example of the elution profiles and retention times used for an SCCP with 60% Cl by weight. Under these conditions, interferences from other co-extracted organochlorine compounds like, PCBs, chlordane, toxaphene, and other organochlorine pesticides, were not observed. One of the big advantages of using this approach is the ability to measure the relative molar amounts of each formula group. This is particularly important because the molecular composition of samples can be very different to that of commercial mixtures that are used as external standards for quantitation purposes [11, 16, 18, 23, 35].

An international inter-laboratory study highlighted the importance of using external standards whose formula group profile closely resembles that of the

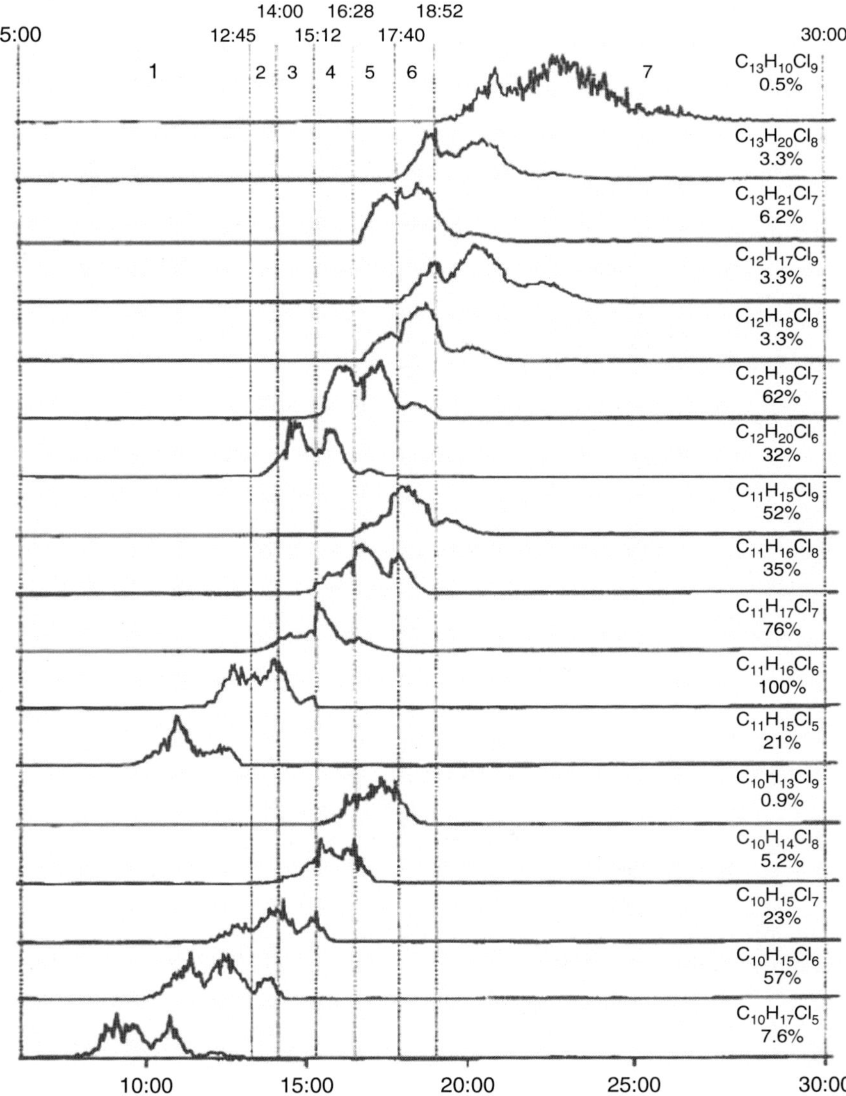

Fig. 8 HRGC/ECNI-HRMS elution profiles of monitored ions in SCCP with 60% Cl by weight aligned to show the retention time windows used [11]

samples [62]. This was further confirmed by Coelhan et al. (2000) who quantitated SCCPs in fish by ECNI low resolution mass spectrometry using several individual CP standards, of varying carbon chain length and chlorine content and also using a technical formulation [63]. Not surprisingly, the results varied by a factor of 10 depending on the standard used. Tomy et al. have shown that it is possible to correct for these differences [11], provided that the formula group profiles are not drastically different between sample and external standard.

Perhaps the biggest advantage to generating formula group profiles is that it enables the fate and behavior of individual formula groups of CPs to be monitored. For example, Houde et al. (2008) were able to examine the trophic transfer and biomagnification of individual formula groups of CPs in two aquatic food webs in Canada [23].

While the ECNI high resolution mass spectrometric method offers unmatched specificity because of the high degree of sophistication and their high cost, these instruments are not available in most analytical laboratories. In light of this, more recent efforts have gone into improving on the early ECNI low resolution mass spectrometric analytical methods.

Reth and Oehme (2004) investigated the mass interferences that can occur in samples under ECNI LRMS conditions [64]. Mass overlap can arise for congeners (in the [M−Cl]$^-$ and [M−HCl]$^{-\cdot}$ ion cluster) with five carbon atoms more and two chlorine atoms less. For example, there is mass overlap between $C_{11}H_{17}{}^{37}Cl^{35}Cl_6$ (m/z 395.9) and $C_{16}H_{29}{}^{35}Cl_5$ (m/z 396.1), which can lead to an overestimation in formula group and/or total CP concentrations. Despite this interference and probable others (e.g., $C_{10}H_{14}Cl_8$ and $C_{15}H_{26}Cl_6$) the authors conclude that careful selection of congener masses, isotope ratios, and retention times can resolve this potential problem [64].

The recognition of potential mass interference in the [M−Cl]$^-$ and [M−HCl]$^{-\cdot}$ ion clusters of CPs themselves and the lower response of the [M−Cl]$^-$ ion relative to $Cl_2{}^{-\cdot}$ and $HCl_2{}^-$ cluster, led Castells et al. (2004) to propose the use of $Cl_2{}^{-\cdot}$ (m/z 70) and $HCl_2{}^-$ (m/z 71) for monitoring purposes [45]. Using GC-ECNI-LRMS, it was concluded that there were no significant mass interferences observed from other organochlorine pesticides, toxaphene, PCBs, and polychlorinated naphthalenes [45]. To achieve this, the authors optimized the ECNI ion-source conditions so that the formation of $Cl_2{}^{-\cdot}$ and $HCl_2{}^-$ ions from co-eluting compounds were negligible. While some organochlorine pesticides gave a small response under the optimized conditions, these compounds were resolved chromatographically and did not affect the quantitation of total SCCPs [45]. Information on individual formula groups is lost using this method.

Negative Ion Chemical Ionization

The underlying difference between ECNI and negative ion chemical ionization (NICI) is that under ECNI conditions there is no ion chemistry between the moderating gas and target analytes in the ion source. Instead, the moderating gas in ECNI is used to thermalize electrons in the ion source [65]. However, in NICI, the reagent gas and target analyte undergo some type of chemical reaction in the ion source.

Methane and argon are the most common moderating gases used for ECNI CP ionization methods. Under these conditions, the common ion cluster monitored is

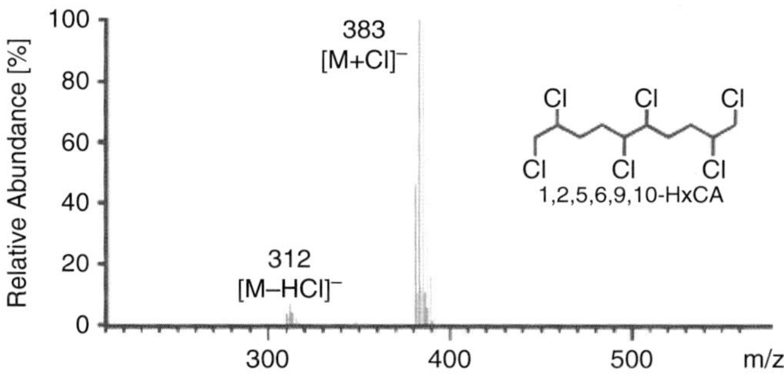

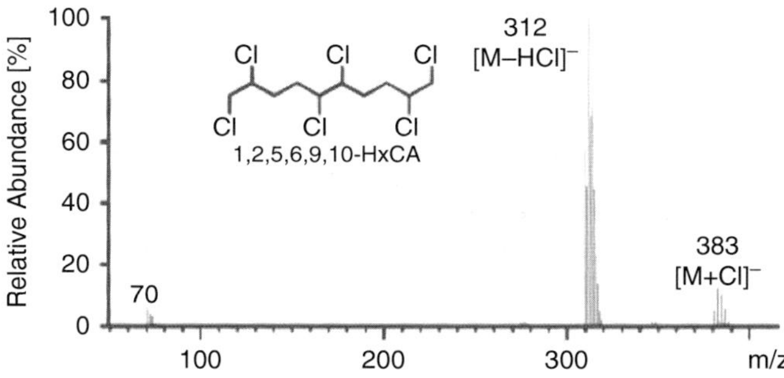

Fig. 9 Comparison of mass spectra of a hexachlorodecane CP congener using CH_4/CH_2Cl_2 (*top panel*) and CH_4-ECNI (*bottom panel*) [49]

from a loss of chloride radical and HCl from the molecular ion. Zencak et al. (2003) proposed the use of a mixture of methane and DCM (80:20) as a reagent gas for NICI [46]. The addition of the CH_2Cl_2 into the reagent mixture induces the exclusive formation of the $[M + Cl]^-$ adduct ion (see Fig. 9). The authors report that the interferences from other CPs and other chlorinated contaminants were significantly reduced and that overall sensitivity improved relative to the conventional methane moderating gas ECNI approach [46]. Taken together, this allowed LRMS to be used in the detection. In addition, the NICI method resulted in similar response factors for congeners with different degrees of chlorination and also enabled the detection of lower chlorinated CP congener, i.e., tri- and tetra-CPs. One major drawback to using the CH_4/CH_2Cl_2 reagent gas is the formation of polymeric material which coats the ion source leading to a reduction in ionization efficiency over time [7].

Electron Ionization

Junk and Meisch (1993) were the first to report on a low resolution EI-MS method for the detection of CPs in environmental samples [62]. By directly introducing a commercial formulation in the ion source of the MS using an insertion probe, the authors were able to select the m/z 105 ion, which corresponded to the molecular fragment, $C_5H_{10}Cl^+$, to be a characteristic ion produced when CPs are ionized under EI conditions [66]. McLafferty (1980) has suggested that this ion is in fact a six-membered cyclic species [67]. The EI fragmentation behavior of CPs has been studied and the stability of some of the proposed cyclic structures assessed using a force-field modeling program and semi-empirical quantum mechanical model [42]. The method by Junk and Meisch was then applied to the analysis of CPs in paving stones from a German metal working industrial plant.

Perhaps the biggest drawback to the use of EI-MS is the large degree of fragmentation it induces on the CP molecule. Numerous fragment ions are formed by consecutive losses of chlorine radicals, by elimination of HCl and by cleavage along the carbon backbone. EI mass spectra of CPs are further characterized by an absence of the molecular ion and by the small abundance of potentially mass-specific high mass fragment ions [42]. Taken together, this suggests that low resolution EI-MS for CP analysis is insufficiently sensitive (large number of fragment ions) and has the potential to suffer from nominal m/z interferences.

More recent developments on EI ion-trap MS have enabled improvements in selectivity and sensitivity for CP analysis [45, 49]. Castells et al. (2004) studied the EI fragmentation of a C_{10}-CP using ion-trap MS and found results that were in general agreement to that of the earlier work of Junk and Meisch (1993) [45, 66]. In an extension of that work, Zencak et al. (2004) reported on fragment ions common to all CPs and proposed selection of specific ion transitions induced using tandem MS (triple quadrupole and ion-trap) [49]. In this approach, the relative response of specific fragment ions of CPs was found to be independent of the chlorine content and carbon chain length. Since all CPs can form these fragment ions independent of their chain length, the method cannot distinguish amongst CPs of varying chain length. While congener and homologue-specific analysis is not possible using this technique, the method is ideally suited when a rapid screening of samples is needed [49].

Metastable Atom Bombardment

Moore et al. (2004) reported on the use of metastable atom bombardment (MAB) ionization and HRMS in the positive ion mode for the detection of CPs [34]. Using an inert gas like argon, MAB ionization generates ions via an electrophilic reaction of a metastable atom with the analyte of interest. Figure 10 shows a comparison of ECNI and positive ion MAB mass spectra.

Using the MAB approach, the two most intense ions in the $[M-Cl]^+$ ion cluster of CP congeners are monitored. A real strength of this method relative to ECNI is the detection of congeners having a small number of chlorine atoms.

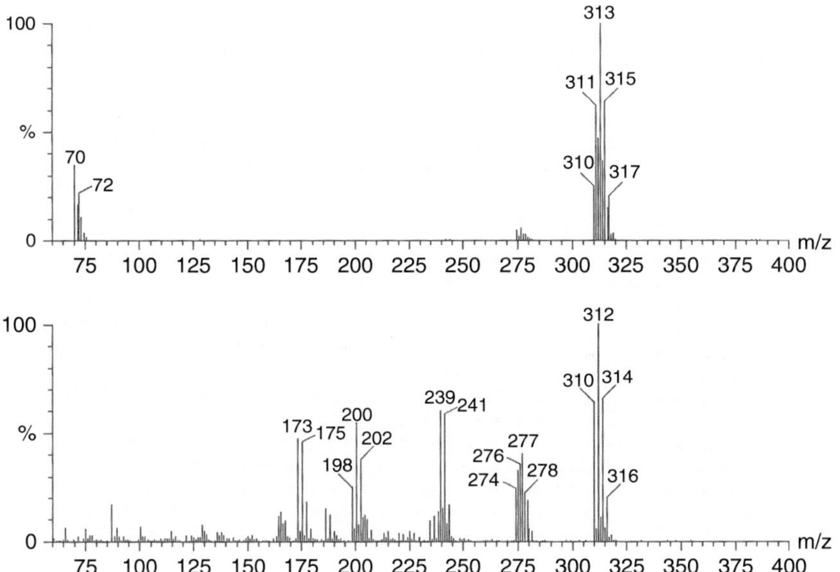

Fig. 10 Mass spectra of hexachlorodecane by ECNI (*top panel*) and metastable atom bombardment (MAB) using Ar (*bottom panel*) [34]

Positive Ion Chemical Ionization

Castells et al. (2004) evaluated the suitability of positive ion chemical ionization (PICI) for the detection of CPs [45]. Figure 11 shows a comparison of an EI and PICI mass spectra of a typical CP. Like EI, mass spectra of CPs in PICI are characterized by the absence of the molecular ion and presence of numerous low-mass fragment ions corresponding to losses of HCl and chloride radical. As such, PICI has not been embraced as a tool for detecting CPs in environmental samples.

3.3 Quantitation

Prior to the advent of capillary GC, PCBs were quantified in environmental samples using technical formulations as external standards. Because different technical formulations of PCBs were available, the approach to quantifying PCBs was to select a formulation whose GC-profile most closely resembled the PCB profile in the sample. This approach has been adopted in the analysis of CPs. Like PCBs, because CPs are known to undergo environmental transformations, it can be challenging to get a technical formulation whose profile closely matches that of the sample.

Early analytical methods that relied on ECD detectors used a similar approach to quantify CPs as was done with PCBs. To circumvent or mitigate the contribution of

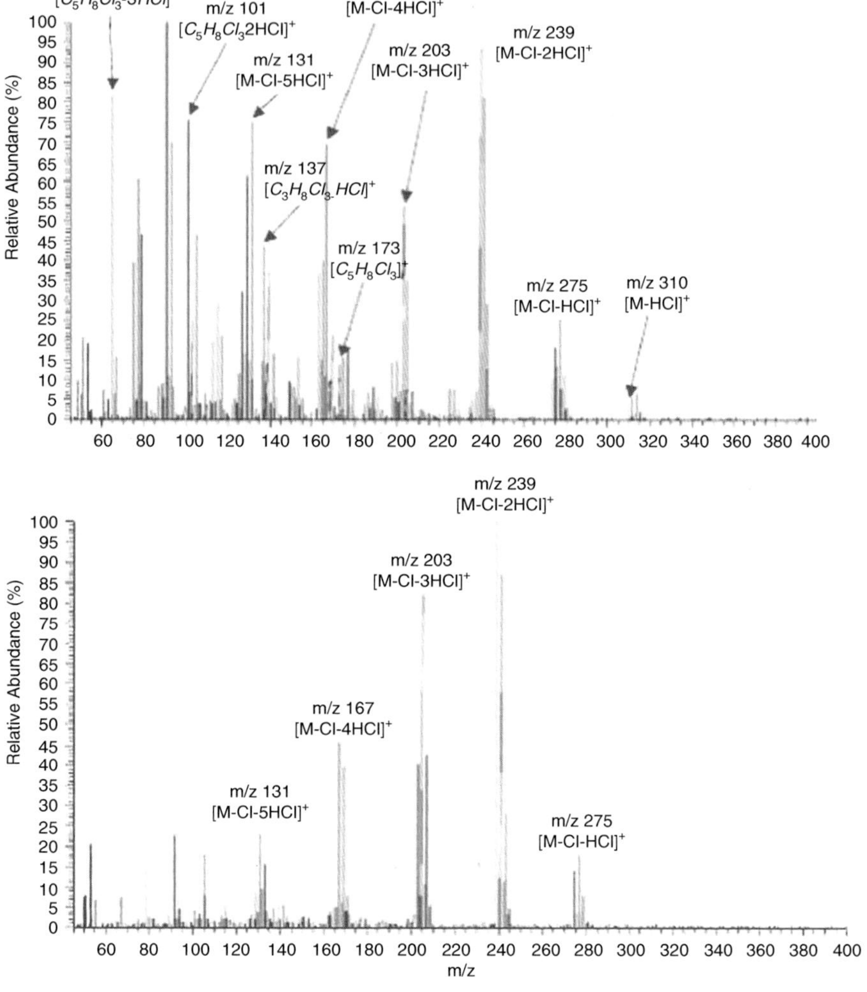

Fig. 11 Electron ionization (EI) (*top panel*) and positive ion chemical ionization (*bottom panel*) mass spectra of hexachlorodecane [45]

interfering peak areas to that of CPs, Walter and Ballschmiter (1991) first proposed the use of a triangulated method for CP quantitation, i.e., constructing a triangle by drawing its base from the start of the CP elution to its end with its apex at the maximum signal (see Fig. 12, [68]).

Coelhan et al. (2000) quantified the extent to which errors can arise when using SCCP standards whose chlorine content differed from that of the sample [63]. Using three standards of varying Cl content, the authors reported that concentration differences of >1,000% were observed when quantifying the same sample.

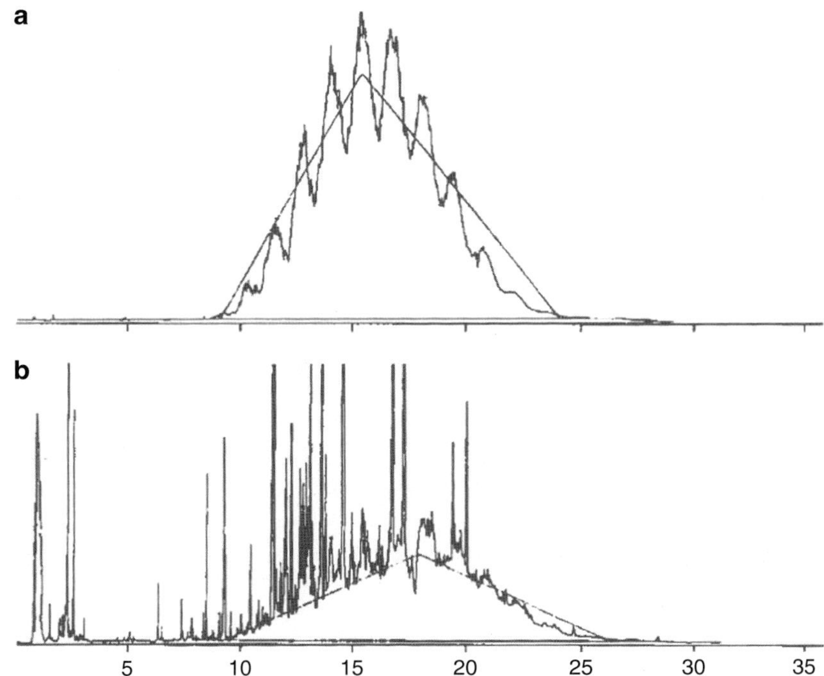

Fig. 12 GC-ECD chromatograms of (**a**) SCCP standard with 60% Cl by weight and (**b**) a fish extract showing the triangles used in calculating the respective areas [62]

Tomy et al. (1997) introduced the numerical correction factor, average molar mass (AMM), to correct for differences in the appearance of the sample and standard [11]. Using the AMM, the authors demonstrated that the error in quantifying one technical mixture using another as an external standard was less than 40% [62].

More recently, Reth et al. (2005) reported on an approach that compensates for the difference in the profiles between sample and standard [69]. The method makes use of the linearity of the ECNI response factor with chlorine content which has only been validated on a quadrupole MS.

Only two interlaboratory studies have been done on SCCPs and none on MCCPs or long-chain CPs (LCCPs) [62, 70]. To reduce as many steps or variables as possible, Pellizzato et al. (2009) supplied six laboratories with an industrial soil extract that was processed in a single laboratory [70]. Identical synthesized standard solutions, now commercially available were use by the laboratories except one laboratory which used an *n*-alkane external standard. Participating laboratories were asked to quantify SCCPs in the supplied extract using an analytical technique of their choice.

All laboratories used GC with a single capillary column, and of the six participating laboratories, five used MS; four in ECNI and one in EI mode. One laboratory used an atomic emission detector (AED). With the exception of a single outlier laboratory, results of the four laboratories that employed MS were in generally good agreement with a range of only 10 ± 1 to 15 ± 2 ng L^{-1}. The one laboratory that employed an AED reported on a value that was ca. four times greater than the mean of the laboratories that used MS detection. Because of the nonselective nature of the AED, the authors felt that this result would improve with better clean-up steps. The one outlier laboratory, which used MS detection, reported on a value that was ca. 200 greater than the other laboratories. It was unclear why there was such a large discrepancy in this one laboratory. The authors concluded that even by eliminating the extraction and clean-up steps, it is still very difficult to obtain comparable SCCP results. Part of the discrepancy was thought to be due to the choice of detection and quantitative approaches used.

Results from an earlier interlaboratory study on SCCPs by Tomy et al. (1999) involving seven laboratories were also met with mixed success [62]. Similar to the Pellizzato et al. (2009) study, participants in the Tomy et al. (1999) study were supplied with an injection ready extract (biota) along with two technical solutions - one with 60% Cl by weight that was used as an external standard and another with 70% Cl by weight that was supplied at an unstated concentration and treated as a 'sample'. A fourth solution containing a purified mixture of UV-chlorinated 1,5,9-decatriene products was also supplied to the participating laboratories at an unstated concentration. All laboratories used single capillary GC coupled with MS operating in the ECNI mode; two laboratories used HRMS run at a resolving power of 10,000–12,000. One laboratory also reported results obtained with an ECD detector.

Surprisingly, measurements made on the synthetic C_{10}-solution whose carbon and chlorine profile differed quite markedly from that of the supplied technical mixtures were closer to the stated concentration (% error range: 2–74%) than measurements on the technical mixture that was treated as a sample (% error: 30–310%). Reasons for this discrepancy were unclear but it was thought that if there were differences in the purity of technical formulations this could make preparation of standard solutions questionable [62].

Overall the analytical precision in measurements made on the fish extract was acceptable. Because two fish extracts were supplied to participating laboratories, two separate measurements were made; the average deviation from the mean (ADM) in the first instance was 10.2 while in the latter it was 16.2. The smaller ADM value was thought to be due to laboratories correcting for interferences from other co-eluting interferences. This was not performed in the latter case.

It is clear from both interlaboratory studies that differences in measurement data on SCCPs (and likely CPs of other chain length) can be notable. The choice of quantitative procedures used by laboratories and also choice of external standard employed (this was certainly the case in the Tomy et al. (1999) study [62]) can contribute to unreliable data. The recent commercial availability of impurity-free synthetic C_{10}–C_{13} solutions should eliminate the uncertainty in preparation of working standard solutions. However, other confounding variables like choice of extraction and clean-up procedures have not been quantitatively assessed and are likely to contribute to the uncertainty in CP data.

4 Conclusion

Over the last 15 years, considerable progress has been made in the analysis of CPs. Much of the analytical efforts have been invested in the analysis of SCCPs with less on MCCPs and relatively little on LCCPs. Mass spectrometric based detectors have clearly facilitated advances in this area.

Perhaps the pioneering work on multidimensional GC offers an enticing glimpse on the future of CP analysis. However, the simplicity of the carbon-skeleton method certainly makes SCCP and analysis of longer chain homologues amenable to more laboratories especially those laboratories new to this area of research.

Perhaps the most pressing issue that needs urgent attention is that of quality assurance. To date, only two inter-laboratory exercises have been conducted; the results from both studies demonstrated that quantitative measurements can be quite varied. Consensus on the choice of a working standard solution, method of quantitation, and certified reference materials will go a long way to ensure that inter-laboratory measurements are more comparable.

With the recent inclusion of SCCP as a candidate POP under international regulatory conventions, it is likely that reliable environmental measurements will be further sought. Addressing the shortcomings that currently exist in the analysis of SCCPs (and other chain length CPs) will go a long way in generating environmental measurement data that can be used by international organizations that regulate the use and release of chemicals.

References

1. Scheer WE (1944) Properties and uses of chlorinated paraffins. Chem Ind 54:203–205
2. Willis B, Crookes MJ, Diment J, Dobson SD (1994) Environmental hazard assessment: chlorinated paraffins. Toxic Substances Division, Department of the Environment, Garston, Watford
3. Kirk-Othmer (1980) Chlorinated paraffins. Kirk-Othmer encyclopedia of chemical technology, 3rd edn. Wiley, New York
4. Svanberg O, Bengtsson BE, Lindén E, Lunde G, Ofstad EB (1978) Chlorinated paraffins – a case of accumulation and toxicity to fish. Ambio 7:64–65
5. Zitko V (1980) Chlorinated paraffins. Springer, New York
6. Mukherjee AB (1990) The use of chlorinated paraffins and their possible effects in the environment. National Board of Water and the Environment, Helsinki, Finland
7. Zencak Z, Oehme M (2006) Recent developments in the analysis of chlorinated paraffins. Trends Analyt Chem 25:310–317
8. Santos FJ, Parera J, Galceran MT (2006) Analysis of polychlorinated *n*-alkanes in environmental samples. Anal Bioanal Chem 386:837–857
9. Eljarrat E, Barceló D (2006) Quantitative analysis of polychlorinated *n*-alkanes in environmental samples. Trends Anal Chem 25:421–434
10. Bayen S, Obbard JP, Thomas GO (2006) Chlorinated paraffins: a review of analysis and environmental occurrence. Environ Int 32:915–929

11. Tomy GT, Stern GA, Muir DCG, Fisk AT, Cymbalisty CD, Westmore JB (1997) Quantifying C_{10}–C_{13} polychloroalkanes in environmental samples by high-resolution gas chromatography/electron capture negative ion high-resolution mass spectrometry. Anal Chem 69:2762–2771
12. Shojania S (1999) The enumeration of isomeric structures for polychlorinated n-alkanes. Chemosphere 38:2125–2141
13. Fredricks PS, Tedder JM (1959) Free-radical substitution in aliphatic compounds. Part II. Halogenation of the n-butyl halides. Chem Soc 144–150
14. Colebourne N, Stern ES (1965) The chlorination of some n-alkanes and alkyl chlorides. J Chem Soc 3599–3605
15. Chambers G, Ubbelohde AR (1955) The effect of molecular structure of paraffins on relative chlorination rates. J Chem Soc 285–295
16. Tomy GT, Stern GA (1999) Analysis of C_{14}–C_{17} polychloro-n-alkanes in environmental matrices by accelerated solvent extraction-high-resolution gas chromatography/electron capture negative ion high-resolution mass spectrometry. Anal Chem 71:4860–4865
17. Coquery M, Morin A, Bécue A, Lepot B (2005) Priority substances of the European water framework directive: analytical challenges in monitoring of water quality. Trends Anal Chem 24:117–125
18. Tomy GT, Stern GA, Lockhart LW, Muir DCG (1999) Occurrence of C_{10}–C_{13} polychlorinated n-alkanes in Canadian midlatitude and Arctic Lake sediments. Environ Sci Technol 33:2858–2863
19. Nicholls CR, Allchin CR, Law RJ (2001) Levels of short and medium chain length polychlorinated n-alkanes in environmental samples from selected industrial areas in England and Wales. Environ Pollut 114:415–430
20. Marvin CH, Painter S, Tomy GT, Stern GA, Braekevelt E, Muir DCG (2003) Spatial and temporal trends in short-chain chlorinated paraffins in Lake Ontario sediment. Environ Sci Technol 37:4561–4568
21. Stejnarová P, Coelhan M, Kostrhounova R, Parlar H, Holoubek I (2005) Analysis of short chain chlorinated paraffins in sediment samples from the Czecg Republic by short-column GC/ECNI-MS. Chemosphere 58:253–262
22. Stevens JL, Northcott GL, Stern GA, Tomy GT, Jones KC (2003) PAHs, PCBs, PCNs, organochlorine pesticides, synthetic musks, and polychlorinated n-alkanes in U.K. sewage sludge: survey results and implications. Environ Sci Technol 37:462–467
23. Houde M, Muir DCG, Tomy GT, Whittle DM, Teixeira C, Moore S (2008) Bioaccumulation and trophic magnification of short- and medium-chain chlorinated paraffins in food webs from Lake Ontario and Lake Michigan. Environ Sci Technol 42:3893–3899
24. Kemmlein S, Hermeneit A, Rotard W (2002) Carbon skeleton analysis of chloroparaffins in sediment, mussels and crabs. Organohalogen Compd 59:279–282
25. Borgen A, Schlabach M, Mariussen E (2003) Screening of chlorinated paraffins in Norway. Organohalogen Compd 60:331–334
26. Castells P, Parera J, Santos FJ, Galceran MT (2008) Occurrence of polychlorinated naphthalenes, polychlorinated biphenyls and short-chain chlorinated paraffins in marine sediments from Barcelona (Spain). Chemosphere 70:1552–1562
27. Pribylová P, Klánová J, Holoubek I (2006) Screening of short- and medium-chain chlorinated paraffins in selected riverine sediments and sludge from the Czech Republic. Environ Pollut 144:248–254
28. Hüttig J, Oehme M (2005) Presence of chlorinated paraffins in sediments from the North and Baltic Seas. Arch Environ Contam Toxicol 49:449–456
29. Parera J, Santos FJ, Galceran MT (2004) Microwave-assisted extraction versus Soxhlet extraction for the analysis of short-chain chlorinated alkanes in sediments. J Chromatogr A 1046:19–26
30. Nilsson M-L, Waldebäck M, Liljegren G, Kylin H, Markides KE (2001) Pressurized fluid extraction (PFE) of chlorinated paraffins from the biodegradable fraction of source separated household waste. Fresenius J Anal Chem 370:913–918

31. Dick TA, Gallagher CP, Tomy GT (2009) Short- and medium-chain chlorinated paraffins in fish, water and soils from the Iqaluit, Nunavut (Canada), area. Int J Environ Waste Manage

32. Castells P, Santos FJ, Galceran MT (2004) Solid-phase extraction versus solid-phase microextraction for the determination of chlorinated paraffins in water using gas chromatography-negative chemical ionisation mass spectrometry. J Chromatogr A 1025:157–162

33. Castells P, Santos FJ, Galceran MT (2003) Solid-phase microextraction for the analysis of short-chain chlorinated paraffins in water samples. J Chromatogr A 984:1–8

34. Moore S, Vromet L, Rondeau B (2004) Comparison of metastable atom bombardment and electron capture negative ionization for the analysis of polychloroalkanes. Chemosphere 54:453–459

35. Tomy GT, Muir DCG, Stern GA, Westmore JB (2000) Levels of C_{10}–C_{13} polychloro-n-alkanes in marine mammals from the Arctic and the St. Lawrence River Estuary. Environ Sci Technol 34:1615–1619

36. Thomas GO, Farrar DG, Braekevelt E, Stern GA, Kalantzi OI, Martin FL, Jones KC (2006) Short and medium chain length chlorinated paraffins in UK human milk-fat. Environ Int 32:34–40

37. Ismail N, Gewurtz S, Pleskach K, Whittle DM, Helm PA, Marvin CH, Tomy GT (2009) Brominated and chlorinated flame retardants in Lake Ontario, Canada, Lake Trout (*Salvelinus namaycush*) between 1979 and 2004 and possible influences of food-web changes. Environ Toxicol Chem 28:910–920

38. Fridén U, Jansson B, Parlar H (2005) Photolytic clean-up of biological sample for gas chromatographic analysis of chlorinated paraffins. Chemosphere 54:1083

39. Iino F, Takasuga T, Senthilkumar K, Nakamura N, Nakanishi J (2005) Risk assessment of short-chain chlorinated paraffins in Japan based on the first market basket study and species sensitivity distributions. Environ Sci Technol 39:859–866

40. Reth M, Zencak Z, Oehme M (2005) First study of congener group patterns and concentrations of short- and medium-chain chlorinated paraffins in fish from the North and Baltic Sea. Chemosphere 58:847–854

41. Rieger R, Ballschmiter K (1995) Semivolatile organic compounds-polychlorinated dibenzo-p-dioxins (PCDD), dibenzofurans (PCDF), biphenyls (PCB), hexachlorobenzene (HCB), 4, 4′-DDE and chlorinated paraffins (CP)-as markers in sewer films. Fresenius J Anal Chem 352:715–724

42. Tomy GT (1997) The mass spectrometric characterization of polychlorinated n-alkanes and the methodology for their analysis in the environment. Ph.D Thesis, University of Manitoba

43. Tomy GT, Fisk AT, Westmore JB, Muir DCG (1998) Environmental Chemistry and Toxicology of Polychlorinated n-alkanes. Rev Environ Contam Toxicol 158:53–128

44. Pellizzato F, Ricci M, Held A, Emons H (2007) Analysis of short-chain chlorinated paraffins: a discussion paper. J Environ Monitor 9:924–930

45. Castells P, Santos FJ, Galceran MT (2004) Analysis of short-chain chlorinated paraffins: a discussion paper. Rapid Commun Mass Spectrom 18:529–536

46. Zencak Z, Reth M, Oehme M (2003) Dichloromethane-enhanced negative ion chemical ionization for the determination of polychlorinated n-alkanes. Anal Chem 75:2487–2492

47. Looser R, Ballschmiter K (1999) Gas chromatographic separation of semivolatile organohalogen compounds on the new stationary phase Optima δ-3. J Chromatogr A 836:271–284

48. Coelhan M (1999) Determination of short-chain polychlorinated paraffins in fish samples by short-column GC/ECNI-MS. Anal Chem 71:4498–4505

49. Zencak Z, Reth M, Oehme M (2004) Determination of total polychlorinated n-alkanes concentration in biota by electron ionization-MS/MS. Anal Chem 76:1957–1962

50. Korytár P, Parera J, Leonards PEG, Santos FJ, de Boer J, Brinkman UAT (2005) Characterization of polychlorinated n-alkanes using comprehensive two-dimensional gas chromatography-electron-capture negative ionisation time-of-flight mass spectrometry. J Chromatogr A 1086:71–82

51. Korytár P, Leonards PEG, de Boer J, Brinkman UAT (2005) Group separation of organohalogenated compounds by means of comprehensive two-dimensional gas chromatography. J Chromatogr A 1086:29–44
52. Korytár P, Parera J, Leonards PEG, de Boer J, Brinkman UAT (2005) Quadrupole mass spectrometer operating in the electron-capture negative ion mode as detector for comprehensive two-dimensional gas chromatography. J Chrom 1067:255–264
53. Koh I-O, Rotard W, Thiemann W (2002) Analysis of chlorinated paraffins in cutting fluids and sealing materials by carbon skeleton reaction gas chromatography. Chemosphere 47:219–227
54. Sistovaris N, Donges U (1987) Gas-chromatographic determination of total polychlorinated aromates and chloro-paraffins following catalytic reduction in the injection port. Fresenius J Anal Chem 326:751–753
55. Randegger-Vollrath A (1998) Determination of chlorinated paraffins in cutting fluids and lubricants. Fresenius J Anal Chem 360:62–68
56. Ballschmiter K (1994) Determination of short and medium chain length chlorinated paraffins in samples of water and sediment from surface water. Abt Anal Chem und Umweltchem
57. Gjøs N, Justavsen KO (1982) Determination of chlorinated paraffins by negative ion chemical ionization mass spectrometry. Anal Chem 54:1316–1318
58. Müller MD, Schmid PP (1984) GC/MS analysis of chlorinated paraffins with negative ion chemical ionization. J High Resolut Chromatogr Chromatogr Commun 7:33–37
59. Jansson B, Andersson R, Asplund LT, Bergman Å, Litzén K, Nylund K, Reutergårdh L, Sellström U, Uvemo U-B, Wahlberg C, Wideqvist U (1991) Multiresidue method for the Gas-Chromatographic Analysis of some Polychlorinated and Polybrominated Pollutants in Biological Samples. Fresenius J Anal Chem 340:439–445
60. Tomy GT, Tittlemier SA, Stern GA, Muir DCG, Westmore JB (1998) Effects of temperature and sample amount on the electron capture negative ion mass spectra of polychloro-n-alkanes. Chemosphere 37:1395–1410
61. Metcalfe-Smith JL, Maguire RJ, Batchelor SP, Bennie DT (1995) Occurrence of chlorinated paraffins in the St. Lawrence River near a manufacturing plant in Cornwall, Ontario. National Water Research Institute, Department of the Environment, Burlington, ON
62. Tomy GT, Westmore JB, Stern GA, Muir DCG, Fisk AT (1999) Interlaboratory study on quantitative methods of analysis of C_{10}–C_{13} Polychloro-n-alkanes. Anal Chem 71:446–451
63. Coelhan M, Saraci M, Parlar H (2000) A comparative study of polychlorinated alkanes as standards for the determination of C_{10}–C_{13} polychlorinated paraffins in fish samples. Chemosphere 40:685–689
64. Reth M, Oehme M (2004) Limitations of low resolution mass spectrometry in the electron capture negative ionization mode for the analysis of short- and medium-chain chlorinated paraffins. Anal Bioanal Chem 378:1741–1747
65. Ong VS, Hites RA (1994) Electron capture mass spectrometry of organic environmental contaminants. Mass Spectrom Rev 13:259–283
66. Junk SA, Meisch HU (1993) Determination of chlorinated paraffins by GC-MS. Fresenius J Anal Chem 347:361–364
67. McLafferty FW (1980) Interpretation of Mass Spectra. University Science Books, Mill Valley, California
68. Walter B, Ballschmiter K (1991) Quantitation of camphechlor/toxaphene in cod-liver oil by integration of the HRGC/ECD-pattern. Fresenius J Anal Chem 340:245–249
69. Reth M, Zencak Z, Oehme M (2005) New quantification procedure for the analysis of chlorinated paraffins using electron capture negative ionization mass spectrometry. J Chromatogr A 1081:225–231
70. Pellizzato F, Ricci M, Held A, Emons H, Böhmer W, Geiss S, Iozza S, Mais S, Petersen M, Lepom P (2009) Laboratory intercomparison study on the analysis of short-chain chlorinated paraffins in an extract of industrial soil. Trends Anal Chem 28:1029–1035
71. Kucklick JR, Helm PA (2006) Advances in the environmental analysis of polychlorinated naphthalenes and toxaphene. Anal Bioanal Chem 386:819–836

Environmental Levels and Fate

Derek Muir

Abstract The environmental levels and fate of chlorinated paraffins (CPs) are understudied compared to many other chlorinated organics with similar molecular formulas such as PCBs and chlorinated pesticides. In this review, the predicted environmental distribution and long range transport (LRT) potential of short, medium, and long chain CPs (SCCPs, MCCPs and LCCPs) homologs were evaluated using a Level III fugacity based model and the OECD LRT tool. Overall persistence (POV) ranged from 503 to 519 d and soils and sediments were predicted to be the main environmental reservoirs for CPs assuming equal emissions to air, water, and soil. Individual congeners are predicted to have quite different reactivity with hydroxyl radicals and congeners with 1,1-dichloro-substitution and adjacent unsubstituted positions on the n-alkane chains having with having in shorter atmospheric oxidation half-lives. CPs have high potential for accumulation in terrestrial food webs based on their phys–chem properties but there is limited confirmation of this. The geographical coverage of environmental levels and trends information on CPs is limited particularly for East Asia, which is now the region with the largest production of CPs. The current lack of environmental levels and fate information, while not the only knowledge gap in the information available on CPs, is a major factor in the uncertainties of recent risk assessments for these substances by regulatory authorities in Europe, Japan, and Canada.

Keywords Bioaccumulation, Distribution, Emission, Fate, Long chain chlorinated paraffins (LCCPs), Medium chain chlorinated paraffins (MCCPs), Persistence, Short chain chlorinated paraffins (SCCPs)

D. Muir
Aquatic Ecosystem Protection Research Division, Environment Canada, Burlington, ON, Canada L7R 4A6
e-mail: derek.muir@ec.gc.ca

J. de Boer (ed.), *Chlorinated Paraffins*, Hdb Env Chem (2010) 10: 107–133,
DOI 10.1007/698_2009_41, © Springer-Verlag Berlin Heidelberg 2010,
Published online: 10 February 2010

Contents

Abbreviations

α-HCH	Alpha-hexachlorocyclohexane
$\delta^{15}N$	Stable isotope ratio of ^{15}N to ^{14}N
AO t1/2	Atmospheric oxidation half-life
AOPWIN	Atmospheric Oxidation Program for Microsoft Windows (part of EPISuite)
CACI-MS	Chlorine addition chemical ionization mass spectrometry
Congener	A structurally unique molecule that is a member of a group of related compounds
CTD	Characteristic travel distance
EBAP	Environmental bioaccumulation potential
ECNI-MS	Electron-capture negative ion mass spectrometry
EPISuite^{TM}	Estimation programs interface (EPI) suite
EUSES	European
HCB	Hexachlorobenzene
Homolog	Having the same molecular formula
HRMS	High resolution mass spectrometry
K_{AW}	Air–water partition coefficient
K_{OA}	Octanol–air partition coefficient
K_{OW}	Octanol–water partition coefficient
LCCPs	Long chain chlorinated paraffins
Level III	Fugacity model based on the work of Mackay et al. – assumes steady state conditions
LRT	Long range transport
LRTAP	Long range transport atmospheric transport
MCCPs	Medium chain chlorinated paraffins
MS/MS	Tandem-mass spectrometry
OECD	Organization for economic cooperation and development
OH	Hydroxyl radical
PCB	Polychlorinated biphenyl
PEC	Predicted environmental concentrations

POV	Overall persistence
QSPR	Quantitative structure property relationship
SCCPs	Short chain chlorinated paraffins
SMILES	Simplified molecular input line entry
SPARC	Sparc performs automated reasoning in chemistry
TE	Transfer efficiency
TMF	Trophic magnification factor
TRI	Toxics release inventory
US EPA	United States Environmental Protection Agency
WWTP	Waste water treatment plant

1 Introduction

Information on environmental levels and fate of chlorinated paraffins (CPs) is crucial to the exposure and risk assessment of these chemicals. Data on environmental levels go back as far as the late 1970s when thin-layer chromatography was used to determine CPs semi-quantitatively in samples from around the UK [1]. A search of the SCOPUS database for "CPs" in the environmental, biological, and physical sciences revealed about 80 published papers and reports that would be broadly classified as related to the determination of environmental levels and fate over the period 1975–2009. By comparison, there are about 5,000 articles on polychlorinated biphenyls (PCBs) using the same search terms.

Although the CPs are very much understudied in terms of environmental measurements compared to other organohalogen compounds, they have nevertheless been reviewed and assessed numerous times as a result of interest from regulators of Existing Chemicals in commerce. Risk assessment reports on CPs are available from the US EPA [2], the European Commission [3–5], and Environment Canada [6]. The Stockholm Convention on Persistent Organic Pollutants has released dossiers on short chain CPs (SCCPs), which include reviews of environmental levels [7, 8]. In Japan, a detailed study of SCCP flows and releases throughout their product life cycle has been conducted [9]. In addition to assessment reports, there are a number of recent comprehensive review articles, which have included environmental levels, trends, and fate [10, 11] as well as earlier reviews [12–14].

As a result, this review, while including recent results, focuses on interpretation of the published results, environmental distribution, and spatial and temporal trends of CPs, and also identifies knowledge gaps.

2 Environmental Entry

As high production volume chemicals, with a large number of industrial and consumer end product uses, CPs are inevitably released into the environment. The EU risk assessments of SCCPs [3, 5] and medium chain CPs (MCCPs) [4] and the Canadian assessment of CPs [6] have considered the releases of CPs into the environment from anthropogenic sources in great detail. The various pathways are summarized in Table 1. Remarkably, there has been little confirmation of these release pathways by actual measurements.

Releases from production sites are thought to be low. The five European production sites for MCCPs are estimated to release about 105 kg/year to waste water treatment plants (WWTPs) or surface waters after treatment. This is a minor fraction of the estimated 44–160 kt annual production at these sites [4]. Emissions to air from manufacturing sites in the EU were considered to be zero. In the USA, C_{10}–C_{13} chloroalkanes (also known as SCCPs) have been on the Toxics Release Inventory (TRI) since 1995. The median estimated emissions for SCCPs to air from manufacturing and use sites over the 13 year period, from 1995 to 2007, was 4.3% of the total of 12–258 t of all on or off site disposal [16]. However, considering that SCCPs were manufactured in the range of >4,545–22,727 t/year, this median value represents 0.25–1.25% of the total produced per year. Median estimated releases to surface water in TRI were 0.24% of all disposal or <0.1% of all manufactured products in the USA. No equivalent estimates are available for MCCPs since only SCCPs are tracked by TRI.

The largest emissions are thought to be from the use and disposal of CPs in metal working fluids. The US EPA estimated these at 10% for SCCPs [2] while the EU assessment estimated releases at 18.5% from oil-based fluids and 31.6% from water-based (emulsifiable) fluids [3]. The other significant source of release is from losses during the service life of products containing CP polymers (PVC, other plastics, paints, sealants, etc.) [3, 4]. These releases, estimated to be overall <1% of the CPs incorporated into the product matrix, are anticipated to be mainly to urban/industrial soil and to wastewater (Table 1).

Recycling of plastics could be a source of MCCPs and LCCPs depending on the extent of dust control at the facility. Yuan et al. [17] found relatively high levels of SCCPs in soils (2.9 µg/g) from an electronic-waste dismantling area in China.

Land-filling is a major disposal route for polymeric products. Minor emissions of CPs from these products could occur for centuries after disposal via leaching from polymers, paints, and sealants [12]. Movement is likely to be limited by sorption of CPs to soils and other organic surfaces.

The fate of CPs in biosolids that are applied to farmland, a common practice in Europe and a growing practice in North America, has not been thoroughly investigated. Nicholls et al. [18] did not detect SCCPs/MCCPs in farm soils amended with sludges containing mg/kg concentrations of CPs. However, worms living in these same soils did contain low mg/kg wet wt. levels of CPs.

CP-containing biosolids may also be incinerated, as would other products, such as CP-containing plastics. The International Programme on Chemical Safety (IPCS) [12] concluded that CPs would degrade in low-temperature combustion due to their thermal instability. Bergman et al. [19] observed that CPs were totally degraded during pyrolysis at 300, 500 and 700°C; however, chlorinated aromatic compounds, including PCBs, were formed during thermal

Table 1 Releases of CPs from production, use, and disposal [3, 4, 6]

Releases from	Product	Pathway	Estimated losses on production[a]	Losses during end use
Production of CPs	SCCPs MCCPs and LCCPs	Wastewaters, air, landfilling	Estimated at 0.3% to wastewaters in Europe; <0.1% in USA via TRI	Not applicable
Formulation and use of metalworking fluids	SCCPs MCCPs and LCCPs	Spills, washoff, volatilization	1–2% due to blending and formulating	10% [2]; 18.5% from oil-based fluids and 31.6% from water-based (emulsifiable) fluids
Production of PVC and other plastics	MCCPs and LCCPs	Spray droplets; volatilization; waste waters	0.3–1.5% from extrusion, calendering, injection molding, coating 0.1–0.2% from open calendering of plastics and plastisol spread coating	0.05%/year to air and waste water
Production and uses of rubber	SCCPs 1–10%; MCCP and LCCP ~15%	Volatilization, abrasion	0.05–0.1%[b]	0.05%/year to air; <0.01% to washoff
Production and uses of textiles and polymeric materials	SCCPs and LCCPs	Washoff of coatings; abrasive losses		0.05% over their lifetime
Production and uses of paints and sealants	MCCPs and LCCPs	Volatilization, washoff, abrasion	0.1% to air and 0.3% to wastewater	0.4%/year to air over 7-year lifetime 0.15%/year to waste water over product lifetime
Releases during plastics recyling	All	Dust particles if chopped or ground	Unknown	Unknown
Sewage biosolids land applications	All	Direct application; runoff; leaching	Unknown	Unknown
Landfilling	All	Leaching	Unknown	Unknown
Incineration	All	PCDD/F formation	Unknown	Unknown

[a]Emission factors apply to annual consumption figure of the CP paraffin in the products
[b]Leaching from rubber/polymers is thought to be much lower than from paints, sealants, and adhesives. No percent loss was given in the assessment [4, 15]

degradation. Pyrolysis of a CP mixture also resulted in the formation of mono and dichlorodibenzofurans; it was however unclear, whether they were formed directly from the CPs themselves or by secondary degradation of PCBs [19]. Tysklind et al. [20] reported formation of PCDDs and PCDFs during combustion of scrap metal containing chlorinated cutting oils. The US EPA's [2] ecological risk assessment of CPs recommended evaluation of the potential for dioxin formation during metalworking and during incineration of waste oil. The Eurochlor [21] website cites unpublished studies indicating no increase in PCDD/Fs during combustion of PVC and other waste streams, both with and without the presence of CPs.

3 Environmental Fate Assessment

Assessments of CPs by Canada and the European Union have utilized simple four compartment (air, water, soil, sediment) fugacity based modeling to predict environmental distribution and exposure concentrations. Environment Canada [6] utilized the EQC model [22] to examine the environmental fate of three MCCPs and one LCCP while the EU assessments of SCCP and MCCPs utilized a Fugacity Level III model [3, 4]. EUSES, the assessment model which implements the methods given in the Technical Guidance Document [23], was also used to calculate Predicted Environmental Concentrations (PECs) of MCCPs. Both the EU assessments used a value of 1×10^{11} h (a default value signifying no degradation) in water, soil, and sediment, which appears to be extraordinarily conservative considering that there is available data on biodegradation rates as well as QSPRs to estimate biodegradation [24]. The EU modeling also did not consider individual SCCP and MCCP homolog groups and simply assigned single values for phys–chem properties and degradation rates to SCCPs and MCCPs. These modeling efforts showed that degradation rates of CPs in soil and sediments were important and also highly uncertain. The updated EU assessment of SCCPs [5] reports a study by Thompson and Noble [25] on biodegradation of two ^{14}C-labeled SCCPs that found mean half-life under aerobic conditions of around 1,630 days in freshwater sediment and around 450 days in marine sediment. Apart from this report, which confirms long environmental half-lives for SCCPs, there has been little progress in developing more accurate biodegradation rates for both MCCPs [15] and SCCPs [7].

The Fugacity Level III and OECD LRT tool models [26] were used to further assess the environmental fate of representative SCCPs, MCCPs, and LCCPs at the level of individual homologs (i.e., $CxH(2x-y+2)Cly$). SPARC software [27] was used to generate log octanol–water partition coefficient (K_{OW}) and the air–water partition coefficient (K_{AW}) required by the models. The predicted log K_{OW}s and log K_{AW}s for 20 CPs are compared with published data in Table 2. The agreement between SPARC predicted log K_{OW}s and log K_{AW}s and the limited published data was excellent for SCCPs, but K_{OW}s were higher than predicted for MCCPs using

Table 2 Comparison of measured and predicted log K_{OW} and log K_{AW}, Fugacity Level III and OECD LRT predictions for selected SCCP, MCCP and LCCP homologs

| CP | Previous values[a] | | SPARC[d] | SPARC | Fugacity level III model[b] | | | | OECD LRT tool[c] | | TE[e] |
| | Used for assessments | | | | Air | Water | Soil | Sediment | POV[e] | CTD[e] | |
	Log K_{OW}	Log K_{AW}	Log K_{OW}	log K_{AW}	%	%	%	%	days	km	%
$C_{10}H_{17}Cl_5$			5.68	−2.89	0.53	3.1	67.2	29.4	503	989	0.48
$C_{10}H_{16}Cl_6$	−		5.77	−3.38	0.23	3.1	62.9	33.8	513	1,107	0.99
$C_{11}H_{19}Cl_5$	6.2	−3.55	6.20	−2.81	0.17	2.1	54.6	43.3	511	791	0.28
$C_{11}H_{18}Cl_6$	6.4		6.27	−3.54	0.17	2.0	51.5	46.4	517	968	0.85
$C_{12}H_{21}Cl_5$	−		6.68	−2.76	0.12	1.4	46.9	51.6	515	623	0.17
$C_{12}H_{20}Cl_6$	6.4–6.8		6.74	−3.61	0.13	1.4	45.6	52.9	518	861	0.75
$C_{13}H_{23}Cl_5$	6.61	−2.77	7.38	−2.82	0.10	1.0	42.3	56.6	518	617	0.22
$C_{13}H_{22}Cl_6$	6.8–7.0		7.5	−3.62	0.11	1.0	42.4	56.4	519	1141	1.63
$C_{14}H_{24}Cl_6$			8.07	−3.22	0.08	1.0	42.3	56.5	519	1,083	1.44
$C_{14}H_{22}Cl_8$			8.37	−4.22	0.07	0.9	46.6	52.4	519	2,608	10.4
$C_{15}H_{28}Cl_6$			8.56	−3.49	0.08	1.0	43.3	55.7	519	1,951	5.64
$C_{15}H_{28}Cl_8$			8.91	−4.35	0.06	0.9	47.7	51.3	519	2,792	12.1
$C_{16}H_{28}Cl_6$			9.08	−3.42	0.06	1.0	45.1	53.9	519	2,388	8.68
$C_{16}H_{26}Cl_8$	−		9.44	−4.28	0.05	0.9	48.5	50.5	519	2,832	12.4
$C_{16}H_{24}Cl_{10}$	7.6	−4.74	9.84	−5.16	0.05	0.9	50.4	48.6	519	2,860	12.7
$C_{17}H_{30}Cl_6$			9.63	−3.70	0.06	1.0	46.2	56.8	519	2,773	11.9
$C_{17}H_{28}Cl_8$			9.99	−4.25	0.05	1.0	49.3	49.7	519	2,851	12.6
$C_{17}H_{27}Cl_9$	7.6	−5.39	10.18	−4.54	0.05	1.0	50.0	48.9	519	2,858	12.6
$C_{18}H_{30}Cl_8$	7.6	−4.54	10.56	−4.35	0.04	1.0	50.0	48.9	519	2,859	12.7
$C_{18}H_{28}Cl_{10}$			11.24	−4.64	0.04	1.1	51.4	47.4	519	2,861	12.7

[a]Log K_{OW} measured or estimated from the equation based on carbon + chlorine number developed by Sijm and Sinnige [28]. Log K_{AW} values measured or estimated from equation of Drouillard et al. [29]

[b]Using the Fugacity Level III model in EPISuite [24] with SPARC estimated Log K_{OW} and Log K_{AW} and EPISuite estimated half-lives in air, water, soil and sediment

[c]Using the OECD LRT tool [26] with SPARC estimated Log K_{OW} and Log K_{AW} and EPISuite estimated half-lives in air, water, and soil

[d]Based on Hilal et al. [27] and available at http://ibmlc2.chem.uga.edu/sparc/

[e]*POV* Overall persistence; *CTD* Characteristic travel distance; *TE* Transfer efficiency

the equations by Sijm and Sinnige [28]. Log K_{OW} and log K_{AW} results predicted by SPARC were not in good agreement with those predicted by KOWWIN and HENRYWIN, which are part of the EPISuite software. Log K_{OW} values in KOW-WIN were 2–4× higher than those predicted by SPARC depending on the actual molecular structure (data not shown). Log K_{AW} values for SCCPs and MCCPs by SPARC ranged from −2.9 to −5.5 while HENRYWIN predicted values of −0.55 to −2.2. Those predicted with SPARC were within the range of measured K_{AW}'s or estimated from the equations developed by Drouillard et al. [29].

EPISuite [24] was used to estimate degradation rates for the CPs. In the case of biodegradation half-lives in water, soil, and sediment, the program assigns a half-life based on the output of BIOWIN3 (an ultimate biodegradation expert survey model; Boethling et al. [30]). BIOWIN3 classifies all CPs as "recalcitrant" for aerobic biodegradation and thus, half-lives for all CPs in the Fugacity Level III model were set at 180 days for water, 360 days for soil, and 1,620 days for sediment. Further rationale for these half-life values is given in the Help section of the EPISuite software (Ver 4.0).

Results of the four compartment Fugacity Level III modeling show that all CPs are mainly associated with soils and sediments under the default conditions of the model, which assume 1,000 kg/h to air, water, and soil compartments. Although no emissions to sediments were assumed, sediments are a key compartment comprising 29–57% of emissions (Table 2). As expected, the % of CPs in the atmospheric compartment is small and declines with increased alkane chain length and chlorine content.

Overall persistence (POV) estimated with the OECD LRT tool ranged from 503 to 519 days. The POV is the residence time at steady state attributable to degradation processes. The Model assumes that substances are transported by the atmosphere and oceans and assumes a land to ocean ratio of 3:7. The narrow range of POVs reflects the fact that degradation rates for soils – the major environmental compartment for CPs – were the same for all homologs. Characteristic travel distances (CTD) and transfer efficiencies (TE), which are dependent on atmospheric half-lives and log K_{AW} values, had a much wider range. The CTD is the point at which deposition flux has reached 37% (1/e) of its initial value. Transfer efficiency is the ratio of the deposition mass flux from air in the remote region to the mass flux in the emitting region. The chlorodecanes and chloroundecanes had TEs <1%. The model predicts that these SCCPs are not efficiently deposited relative to MCCPs, reflecting shorter CTDs. The shorter CTDs for SCCPs seems counter intuitive since atmospheric oxidation half-lives (AO t1/2) of SCCPs are predicted (using EPISuite AOPWIN) to be generally longer (1.5–2.2 day) than those of MCCPs (1.0–1.5 day). However the model predicts much greater removal of SCCPs by atmospheric oxidation due to the higher proportion in the atmosphere (as illustrated by the Fugacity model results in Table 2).

The most useful aspect of the modeling assessment is comparison with other compounds. In Fig. 1, CTD and TE results for the CPs and for a series of reference chemicals, are plotted against POV values. The reference chemicals are pesticides, PCB homologs, and other organics in the OECD LRT database. The SCCPs and

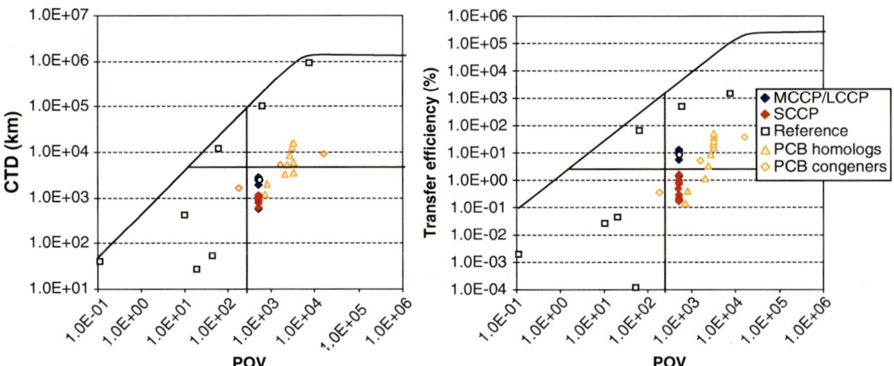

Fig. 1 Results for characteristic travel distance (CTD) and Transfer efficiency (TE) for 20 SCCP–MCCP–LCCP homologs in Table 2 are compared with reference compounds (α-HCH, aldrin, atrazine, biphenyl, CCl_4, HCB, p-cresol), PCB congeners (CB 28, 101, 180), and PCB homologs (Cl_1 to Cl_{10}). The curved line is the "volatility line" representing a physical limit on transport potential in the atmosphere defined by a hypothetical substance that partitions entirely into air [31]

MCCPs can be seen to be intermediate in their positioning on the plots in Fig. 1 with shorter POVs than almost all PCBs (except PCB 28), but longer than all reference chemicals except CCl_4 and HCB, which are positioned in the upper quadrant defined by the line at POV = 195 days (for α-HCH), CTD = 5,089 km (for PCB-180) and TE = 0.00065% in each graphic. Most of the PCBs are also in this quadrant due to long CTD and high TE. These quadrants were suggested by Klasmeier et al. [31] as boundaries defining high and low long range transport and deposition. By these criteria, the SCCPs and MCCPs fall mainly below the "LRTAP" boundary although they have high POVs due to the long predicted persistence times in water and soil.

To this point, only the predicted environmental fate of the CP homologs has been discussed. However, the major homologs within the SCCP/MCCP/LCCP categories consist of thousands of individual congeners [32]. With the exception of a small number of congeners or congener mixtures that have been synthesized [29, 33] it is currently not possible to assess CPs at this level of detail. However, SPARC will calculate the physical properties of structural isomers and the EPISuite AOPWIN program can generate AO t1/2s that vary with molecular structure. Therefore in Table 3, the AO t1/2s, log K_{OW}, log K_{AW}, and log K_{OA} of 10 hexachlorodecanes are compared as a case study. Hexachlorodecanes are major components of SCCPs, comprising the majority of the chlorodecanes in two SCCP formulations analyzed by Tomy et al. [34] and Hüttig and Oehme [35]. AOt1/2s for the hexachlorodecanes mono-substituted at the 1, 2, and/or 9, 10 positions are predicted to be very similar, ranging from 2.2 to 2.9 days for the seven examples in Table 3. However, the two congeners lacking a terminal chlorine are predicted to have longer AO t1/2s (3.5–4.0 day). A 1,1-dichloro- substituted congener is predicted to have a shorter AO t1/2 (1.9 day), presumably due to the reactivity of the terminal carbon. Also, congeners with 4 consecutive unsubstituted carbons had the next lowest AO t1/2 (2.2 d). The greater in reactivity of the 1,1-disubstituted

Table 3 Predicted atmospheric oxidation half-lives, log K_{OW}, log K_{AW} and log K_{OA} for selected hexachlorodecane congeners

SMILES notation[a]	Structure	AO[b] t1/2 (day)	Log K_{OW}	Log K_{AW}	Log K_{OA}
ClCCC(Cl)C(Cl)C(Cl) CCC(Cl)CCCl	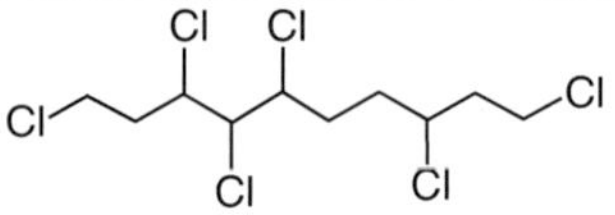	2.8	5.83	−3.79	9.62
ClCCCCCC(Cl)C(Cl)C (Cl)C(Cl)CCl	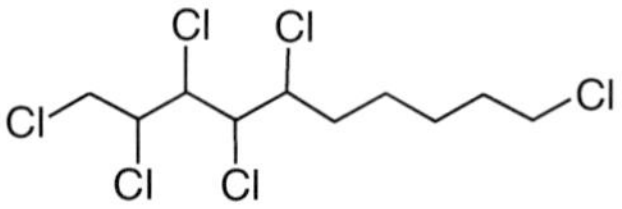	2.2	5.82	−3.81	9.63
ClCCC(Cl)CC(Cl)C(Cl) CC(Cl)CCCl	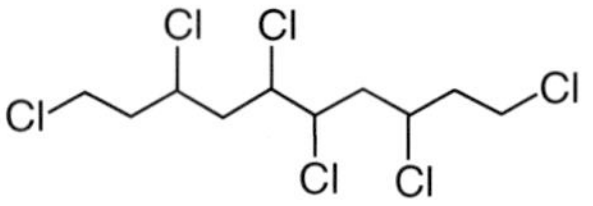	2.5	5.85	−3.57	9.42
ClCC(Cl)CC(Cl)CC(Cl) CCC(Cl)CCl	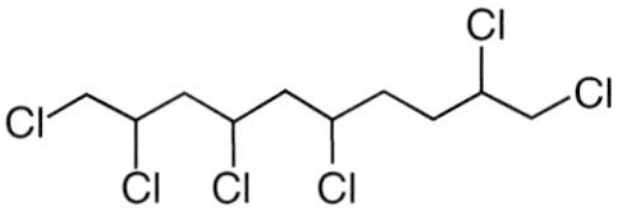	2.5	5.93	−3.35	9.28
ClCCC(Cl)CC(Cl)C(Cl) CCC(Cl)CCl	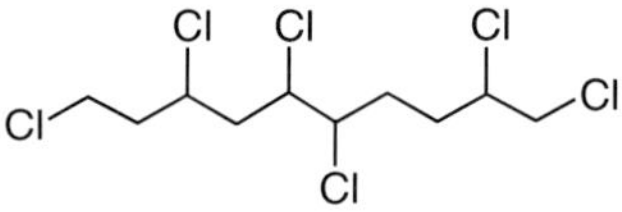	2.9	5.89	−3.49	9.38
ClCCC(Cl)CC(Cl)CC (Cl)CC(Cl)CCl	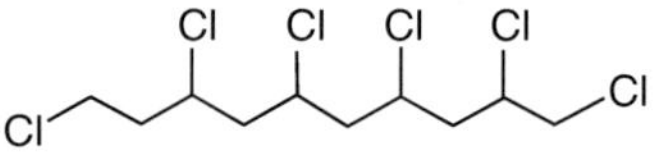	2.3	5.9	−3.37	9.27
ClCC(Cl)C(Cl)CC(Cl) CCCC(Cl)CCl	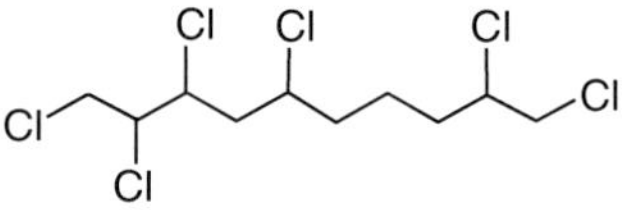	2.4	5.93	−3.45	9.38
ClCC(Cl)CCC(Cl)C(Cl) CC(Cl)C(Cl)C	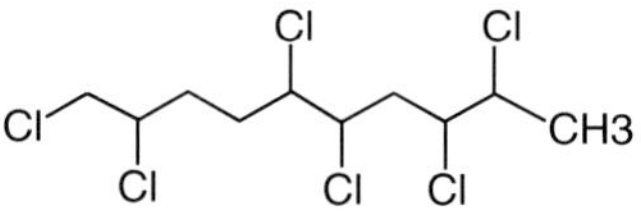	4.0	5.84	−3.31	9.15
ClCC(Cl)C(Cl)CC(Cl)C (Cl)CC(Cl)CC	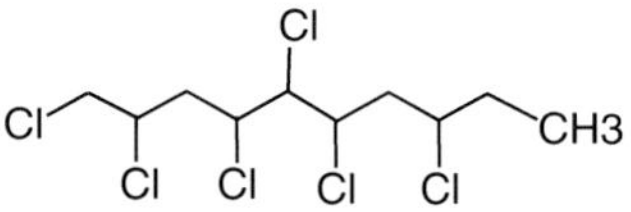	3.5	5.76	−3.38	9.14
ClC(Cl)CC(Cl)CC(Cl) CCCC(Cl)CCl	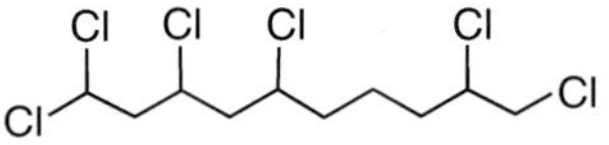	1.9	6.17	−3.04	9.21

[a]Simplified molecular input line entry specification [24]
[b]Atmospheric oxidation half-life (days) predicted from AOPWIN (EPISuite 2008; Ver 4.0)

congener and the congener with adjacent unsubstituted positions illustrates the potential shifts in patterns of the hexachlorodecanes and related SCCPs, with similar % chlorine, as these compounds are oxidized in air or surface waters. El-Morsi et al.

[36] found that 1,2,9,10-tetrachlorodecane underwent rapid aqueous photoxidation by OH-radicals generated from H_2O_2. However, there appear to be no other studies of the gaseous phase photooxidation of CPs.

Log K_{OW}s, log K_{AW}s, and log K_{OA}s for the 10 congeners vary over a relatively narrow range (+/−0.4 log units) compared to the AOt1/2s. The 1,1-dichloro-substituted congener has the highest predicted log K_{OW} (6.17) and the lowest log K_{AW} (−3.04) of the group analyzed. As Tomy et al. [34] have noted, CPs are likely to have no more than one chlorine atom bound to any carbon atom because in free-radical chlorination used for CP synthesis, a second chlorine atom would not readily substitute for hydrogen at a carbon already bound to chlorine. An exception to this might be at a terminal carbon, which is why the 1,1-substituted isomer was included in Table 3 for illustrative purposes. The modeling results suggest that reactions with OH, as well as metabolic oxidation by cytochrome mixed function oxidase enzymes are more likely to influence the actual pattern of CPs observed in environmental samples than partitioning processes.

4 Levels and Trends by Environmental Compartment

4.1 Sediments and Water

Measurements of CPs in sediments constitute the largest set of data on the environmental levels of CPs. The early literature was reviewed by Tomy et al. [13] and IPCS [12]. Recent assessments of SCCPs and MCCPs [4–6, 15] have reviewed the available data for sediment and water to about the mid-2000s. Recently Feo et al. [11] thoroughly reviewed the reports on CPs in sediments and water. They noted that SCCPs and MCCPs were detected in almost all samples but with a wide range of concentrations. This variation is due to a variety of factors that influence partitioning of hydrophobic organics from the dissolved phase to suspended particles and deposition in sediment, and below, it is discussed further in terms of the spatial and temporal trends of SCCPs and MCCPs reported since 2000.

Spatial trends: As might be expected for chemicals used in industrial applications such as metal working, as well as in consumer products, proximity to industrial and urban areas explains much of the spatial variation seen in CPs in sediments. Data from United Kingdom showed heavy contamination of freshwater and estuarine sediments collected near specific industrial or WWTP sources by SCCPs and MCCPs with concentrations up to 65 μg/g [18] while coastal sediments along the Catalonia coast near Barcelona (Spain) and under the influence of wastewater treatment plant (WWTP) effluents and sewage sludges had up to 2 μg/g SCCPs [37]. Hüttig and Oehme [38] did not observe a gradient of SCCPs and MCCPs in the North Sea away from the Elbe river estuary. They also noted significantly different CP homologue patterns from one year to the next in sediments sampled at the same site in two consecutive years. Normalizing with total organic carbon (TOC) reduced the spatial variation in the CP concentrations.

Marvin et al. [39] found wide variation of SCCPs (<5–410 ng/g) in 25 sediment samples representing the top 3 cm from widely spaced sites in Lake Ontario with the highest concentrations in depositional areas (147–420 ng/g). The highest concentrations were found in the western basin of Lake Ontario, which is relatively close to major urban areas and WWTPs. Similarly, much higher SCCP levels (135 ng/g) were found in surface sediments in the south basin of Lake Winnipeg, which receives WWTP effluent from upstream sources, compared to the remote northern basin of the lake (8 ng/g). Přibylová et al. [40] found SCCPs in river sediments in the Czech Republic ranging from <~2 to 347 ng/g and MCCPs ranging from <~2 to 5,575 ng/g with the highest levels near industrial sites where CPs were probably used for metal working. Iino et al. [41] found higher SCCP concentrations in downstream samples of two rivers in Japan (Kasaibashi and Denpo Ohashi; 484 and 424 ng/g dw respectively) than the upper stream sites in the same systems (211 and 4.9 ng/g) and noted a relationship to sediment TOC.

Background sites such as lakes in the Canadian arctic [42, 43] also had detectable concentrations, typically in the low ng/g range for total SCCPs while Přibylová et al. [40] found SCCPs and MCCPs in river sediments at some background sites in rural Czech Republic. Surprisingly, marine sediments in the Canadian Arctic Archipelago had SCCP concentrations ranging from 4.5–77 ng/g (dw) with the highest levels on the eastern side of the archipelago [44].

CP homolog patterns in sediments and water: While total SCCP/MCCPs or sum of chain length groups (ΣC_{10}, ΣC_{11}, etc,) have generally been reported, recent studies have included individual CP homologs (i.e., same chain length and chlorine content). This permits greater insights into the environmental fate of CPs. However, it should be noted that the homolog pattern can change depending on the analytical method. Hüttig and Oehme [35] found that the main "congeners" (actually homologs) of a given C-chain had one Cl more when detected by GC-electron capture negative ion (ECNI)-MS compared to chlorine addition chemical ionization (CACI) MS. Thus $C_{13}H_{24}Cl_4$ was the main SCCP homolog and $C_{14}H_{26}Cl_4$ the main MCCP homolog in nearly all marine sediment samples analyzed by Huttig and Oehme when CACI-MS was used for detection while the corresponding Cl_5 homolog predominated by ECNI-MS. Similarly, using metastable atom bombardment ionization (MAB) and high resolution mass spectrometry, Moore et al. [45] noted that chlorinated alkanes with three and four Cl atoms accounted for between 10 and 46% of total SCCPs in river water. Nevertheless, most studies have used ECNI-MS due to its greater sensitivity (at least for chlorinated alkanes with >4 Cls) and ease of application. Iino et al. [41] reported Cl_4 containing C_{10}–C_{13} SCCPs using ECNI-HRMS.

Figure 2 compares the results for SCCP homologs in sediments and water from three studies [35, 40, 45] that have analyzed a large number of SCCP homolog groups ranging from Cl_4 to Cl_{11} SCCPs. Moore et al. [45] detected Cl_3 chlorodecanes and undecanes in suspended solids and dissolved phase implying dechlorination products were present since this homolog was not present in a 60% Cl SCCP commercial product. Hüttig and Oehme [35] also noted the relatively high proportion of Cl_4 homologs (~10%) in a 55.5% Cl SCCP product but did not determined

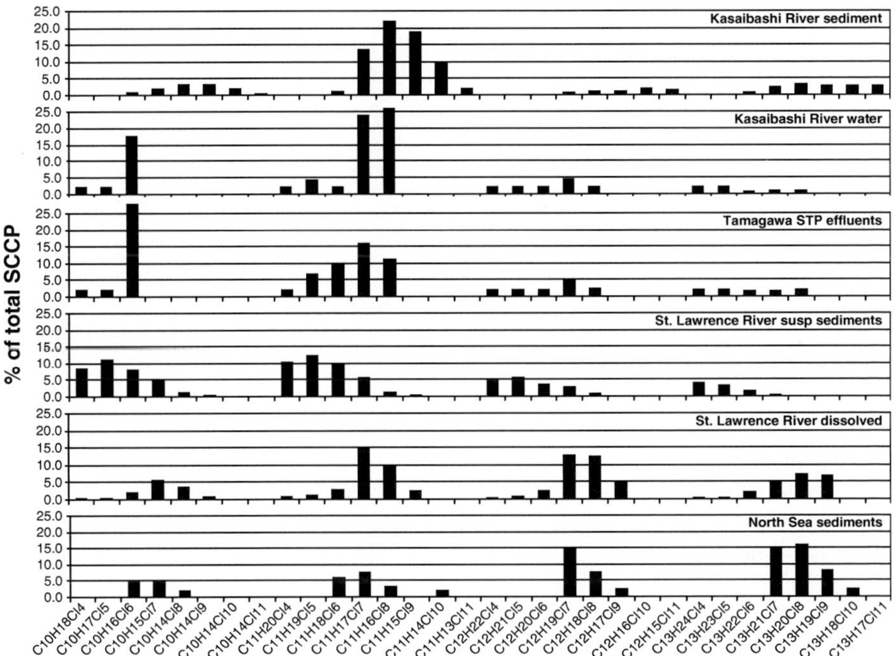

Fig. 2 Comparison of SCCP homolog group patterns in the Kasaibashi River (Japan) and nearby Tamagawa (Tokyo) STP effluents [41], St Lawrence river suspended (susp) sediments and dissolved phase [45] and in North Sea sediments [35]. Results from Hüttig and Oehme [35] are combined for their sample 1 (North Sea) and 8 (Baltic Sea)

Cl_3 homologs. The patterns of SCCPs are relatively similar in the Japanese and European sediments, with $C_{11}Cl_7$ and $C_{11}Cl_8$ predominating. The St Lawrence River's suspended sediments had $C_{11}Cl_5$ and $C_{11}Cl_6$ predominating. However, Marvin et al. [39] also found relatively high proportions of $C_{11}Cl_7$ in Lake Ontario sediments, so the lower proportion in suspended sediments could reflect differences between depositional and suspended phases. Hüttig and Oehme [35] also reported individual MCCP homologs in the North Sea and southwest Baltic Sea sediments. They demonstrated that $C_{14}Cl_6$ and $C_{14}Cl_7$ homologs predominated in sediment (by ECNI-MS) although $C_{14}Cl_4$ and $Cl_{14}Cl_5$ were prominent when using CACI-MS. Their results suggest that the MCCP technical product was dechlorinating in sediment. Few other data exist with which to compare patterns for individual MCCP homologs in sediments.

CP historical profiles in sediment. The spatial trends in sediments discussed above are confounded by varying rates of sedimentation so that some samples may represent the entire history of deposition of CPs while others may be recent deposition only. This is particularly clear in the study by Marvin et al. [39] in Lake Ontario. Sediment cores, dated using ^{210}Pb and ^{137}Cs, can be used to derive fluxes (ng m^{-2} year^{-1}) and are thus more appropriate for comparing locations and also for assessing historical inputs. Unfortunately the geographical range of dated sediments that have been analyzed for CPs is limited. As of September 2009, there was detailed work only for one lake in Europe (Lake Thun in Switzerland [46]) and from Lake Ontario [39] and other lakes in Canada [42–44].

Figure 3 compares the historical profiles of SCCPs in dated sediment cores from six lakes. Two lakes (Winnipeg and Ontario) are downstream of, or immediate adjacent to, urban areas and receive direct WWTP effluents [39, 48] while four others vary from low population density (Thun, Switzerland) [46] and Nipigon, Canada [42]) to completely remote (DV09 and Yaya, Canada) [42, 43]. When normalized to the maximum concentration to allow comparison of historical profiles, most cores show the increase in SCCP production that occurred in the 1970s and 1980s in North America and Western Europe. Rising concentrations occur in the 1960s and 1970s in the 3 cores downstream of urban areas which are generally earlier than in the four other lakes (Fig. 3). This may be due to direct releases of SCCPs from metal working which is a major activity (e.g., vehicle production and associates parts industries), upstream of this sites. These cores also show earlier elevations possibly associated with the increased production reported during World War II. Following the US EPA assessment of SCCPs in the early 1990s [2], the CP producers in North America and Europe implemented guidelines to reduce SCCP emissions from metal working [49, 50]. This reduction is not clearly apparent from

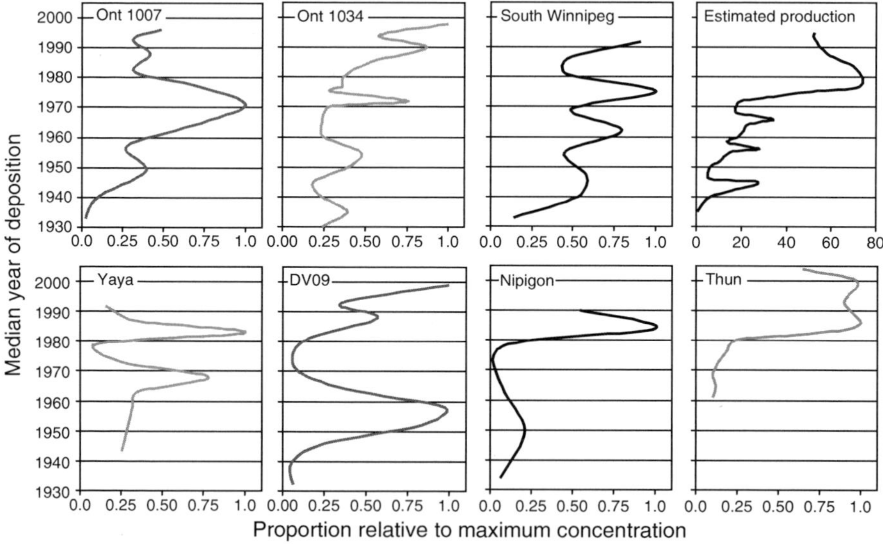

Fig. 3 Comparison of historical profiles of SCCPs in dated sediment cores from Lakes Thun (Switzerland) [46], Ontario (Ont1007 and Ont1034) [39], other Canadian lakes [41], and DV09 a high latitude lake in the Canadian arctic [42]. Results are normalized to the maximum concentration reported to allow comparison. Profiles are compared to North American and European production of SCCPs (tons/year) estimated in Muir et al. [47]

the historical profiles which suggest that the reduced emissions may have occurred earlier, i.e., in 1980s. Some cores in North America even in remote areas (DV09, Nipigon) show increased SCCP inputs in horizons dated to WWII or the early 1950s suggesting the use pattern at the time was resulting in atmospheric emissions. The Lake Thun core, the only one from Europe, was not analyzed far back enough historically to confirm a WWII signal. This core also shows a CP maximum in the mid-1980ss, about 5–10 years later than the estimated maximum SCCP production in Europe and North America.

SCCPs are clearly detectable in sediment horizons that are older than 50 years, indicating their prolonged persistence under anaerobic conditions once buried in lake bottoms. However, the pattern of homolog groups does show changes over time as illustrated in Fig. 4 which shows the relative increase or decrease in the percentage in the surface slice and in sediment horizons dated to pre-1950 (negative % indicates lower % in the deep sediment). Both sites, with direct WWTP and upstream industrial sources (Winnipeg and Ontario), show declining proportions of pentachlorodecanes. However, among the undecanes and dodecanes, only the higher chlorinated homologs (Cl_7–Cl_8) show a reduced proportion at both sites. The remote site (DV09) shows quite different shifts of SCCP homologs with higher Cl_5 and Cl_6 decanes and undecanes in the pre-1950 sediment than in recent horizons.

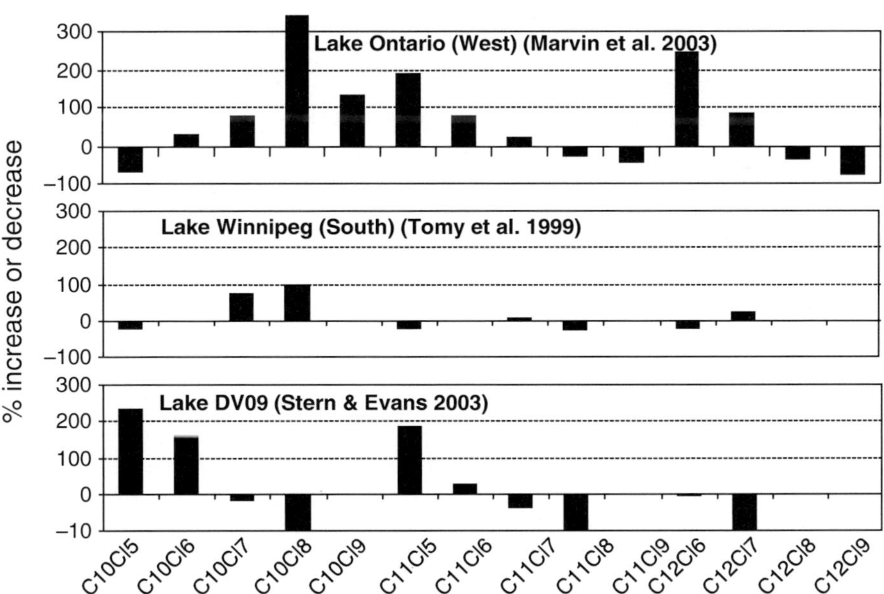

Fig. 4 Relative change (%) of major SCCP homolog groups in three lake sediment cores calculated from the percentage in the surface slice and the percentage in the sediment horizon dated to pre-1950. A negative number indicates higher proportion in surface than the deep slice

Whether these shifts are due to microbial reductive dechlorination which has been observed for toxaphene congeners [51] and for PCBs in sites with relatively high concentrations [52], or is due to shifting SCCP sources is difficult to determine without actual analyses of SCCP enantiomers. If this could be done, it would help to assign sediment half-lives as has been done with PCBs [52]. Also, if techniques which detect the tri- and tetrachloro C_{10}-C_{16} alkanes could be employed for analysis of sediment cores, e.g., CACI-MS or MAB-MS, it should be possible to infer the extent of dechlorination (as has been done with toxaphene congeners [51] and for DDT to DDD in sediments [53] since, as noted above, there is evidence from analysis of surface and suspended sediments of greater proportions of Cl_4 and Cl_5 homologs than are present in technical products [35].

4.2   Fish and Aquatic Food Webs

Analysis of CPs in aquatic biota constitutes the next largest environmental dataset for SCCPs and, to a much lesser extent, for MCCPs/LCCPs (Feo et al. [11]). The early literature on levels in biota was reviewed by Tomy et al. [13] and IPCS [12] as well as in the US, EU, and Canadian assessments of SCCPs and MCCPs [2; 3]. More recent assessments of SCCPs and MCCPs [4–6, 15] have thoroughly reviewed the available data to about the mid-2000s. Therefore, this review will essentially focus on studies that included the patterns of SCCP and MCCP homologs, thus permitting discussion of factors that may influence these patterns.

Spatial trends and homolog patterns: Only a very limited number of studies have reported the homolog patterns in various aquatic biota; the majority have reported only total SCCPs and MCCPs [11]. Reth et al. [54, 55] reported SCCP–MCCP concentrations and homolog patterns in a range of fish species from the North Sea, the Baltic Sea, the North Atlantic, and Lake Ellasjøen, a lake impacted by seabird guano, on Bjørnøya in the Arctic Ocean 500 km north of Norway. Although sample sizes were limited, the SCCP and MCCP concentrations were higher in Atlantic cod (*Gadus morhua*) from the Baltic (19–143 ng/g ww and 25–106 ng/g ww, respectively) than in the same species from Iceland (11–70 ng/g ww and 7–47 ng/g ww, respectively). North Sea dab (*Limanda limanda*) had higher concentrations of SCCP–MCCPs than cod or flounder (*Platichthys flesus*).

SCCP–MCCP concentrations in arctic char from Ellasjøen (11–27 and 13–43 ng/g ww in liver, respectively) were much lower than total PCBs which averaged from about 500–1,000 ng/g ww in separate collections [56]. Two species of seabirds, little auks (*Alle alle*) and kittiwakes (*Rissa tridactyla*) collected at Bjørnøya had a similar range of SCCP–MCCPs concentrations in liver and muscle as the char liver. There were relatively similar proportions of C_{10}, C_{11}, C_{12}, and C_{13} SCCP chain length groups in char, and the two seabirds possibly reflecting the fact that seabird guano is considered the major source of SCCPs to Ellasjøen [57].

Dick et al. [58] analyzed anadromous char (*Salvelinus alpinus*) from Iqaluit in the Canadian arctic and found SCCPs and MCCPs in char muscle averaging

7.8 ± 17 and 6.8 ± 11 ng/g lipid weight, respectively. Although the anadromous char were collected near an abandoned military site that was leaking SCCPs and MCCPs, the concentrations were nevertheless 54 and 150-times lower than SCCP and MCCP concentrations found by Reth et al. [55] in char from Ellasjøen. The difference may be mainly attributed to the elevated inputs of all bioaccumulative organochlorines in Ellasjøen due to bird guano. In addition, the proximity of Bjørnøya to Northern Europe means that Ellasjøen could receive greater inputs due to long range transport and this could be a contributing factor to the high SCCP levels.

Houde et al. [59] found higher concentrations of SCCPs in lake trout (*Salvelinus namaycush*) from northern Lake Michigan (123 ± 35 ng/g ww) than Lake Ontario (34 ± 37 ng/g ww); however, MCCPs were higher in Lake Ontario lake trout (24 ± 26 vs. 5.6 ± 4.8 ng/g). Similar to observations for arctic char in Lake Ellasjøen, SCCP–MCCPs in Lake Ontario lake trout were much lower (50×) than PCBs in the same fish [47].

Figure 5 compares the SCCP–MCCP homolog pattern in Atlantic cod collected north of Iceland with the same species in the Baltic Sea (all reported by Reth et al. [54, 55]) and with flounder, lake trout [59], arctic char [58], and a mixed set of fishes from Japan (SCCPs only) reported by Iino et al. [41]. These patterns help in the assessment of CP sources and bioaccumulation potential.

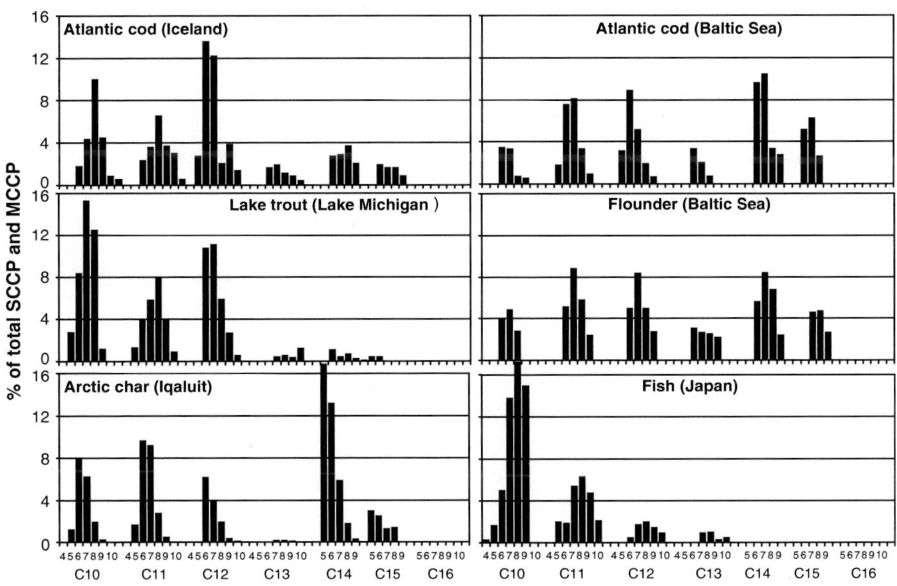

Fig. 5 Comparison of SCCP and MCCP homolog group patterns in lake trout [59], Atlantic cod liver from Iceland [55], Arctic char from near Iqaluit (Canada), cod and flounder from the Baltic [54], and market fish from Japan [41]. The results from Japan are for SCCPs only. Proportions from Reth et al. [54, 55] were estimated from graphical results. Numbers of chlorines for each homolog are shown below each bar for C_{10}–C_{16} carbon chains

A likely indicator of direct urban WWTP and industrial emissions influencing concentrations in fish is the presence of higher proportions of MCCPs. This is seen in cod and flounder from the Baltic Sea and arctic char from Iqaluit. At Iqaluit, MCCPs were higher than SCCPs in stream water that flowed into Cumberland Sound near where the arctic char were collected. In the case of the Baltic Sea, Hüttig and Oehme [35, 38] found about 2-fold higher MCCPs than SCCPs in surface sediments. Fish from more remote locations (Atlantic cod collected near Iceland and lake trout from northern Lake Michigan) have low proportions of MCCPs. MCCPs are generally expected to be more particle bound and therefore less mobile in the aquatic environment.

Within the SCCP group of chloro-n-alkanes, there is no clear trend to higher proportions of less chlorinated, more volatile homologs in fish samples from remote areas. In most fish samples Cl_5, Cl_6, and Cl_7 n-alkanes predominate for both SCCPs and MCCPs. These homologs predominate in commercial CP products [35, 38]. They also have optimum molecular size and hydrophobicity for bioaccumulation [60, 61]. The presence of Cl_4-alkanes in fish was detected by Iino et al. [41] while Reth et al. [54, 55] did not report them in Baltic, North Sea, or North Atlantic fish. Whether the presence of Cl_4-n-alkanes indicates uptake of dechlorinated products or exposure to commercial products with these homologs is unclear. Overall, the pattern of SCCP homologs in fish and in sediments and effluent samples from Japan differed markedly from that found in Canada and northern Europe (Figs. 2 and 5). Presumably this reflects the differences in commercial products; however, there are as yet no published reports on homolog patterns in SCCPs manufactured or sold in Japan or other Asian countries.

Food web accumulation: Given that sediment is the major environmental compartment for CPs, the bioavailability from sediment to bottom feeding invertebrates and fish is a key pathway for exposure of upper trophic level organisms in aquatic food webs. Laboratory studies with oligochaetes (*Lumbriculus variegatus*) have demonstrated that SCCP–MCCPs are readily accumulated from sediments [62]. Houde et al. [59] noted that the relative contribution of SCCP carbon chain groups in slimy sculpin (*Cottus cognatus*) was very similar to results reported for sediment samples collected in Lake Ontario, supporting the hypothesis that sediment is a source of CP contamination in sculpin.

This sediment-biota connection is further explored in Fig. 6 which compares SCCP homologs in sediment (black bars) versus bottom feeding fish (flounder and sculpin) from the same area. Unfortunately insufficient data was available to compare MCCPs in the same samples.

The pattern of SCCPs in the Baltic Sea sediment (Site 8 in Hüttig and Oehme [35]) is unusual in that that there are high percentages of C_{13}-chloro-n-alkanes. The $C_{13}Cl_7$ and $C_{13}Cl_8$ pattern is not observed in flounder from the same area (OS3 in Reth et al. [54], however, the proportions of C_{10}–C_{12} homologs in sediments are reflected in flounder. Similarly in Lake Ontario, the C_{11} and C_{12} homolog pattern is reflected in sculpin (Fig. 6) while the relatively high proportions of $C_{13}Cl_7$ and $C_{13}Cl_8$ in sediment are not. Also $C_{10}Cl_5$ and $C_{10}Cl_6$ are underrepresented in sculpin compared to sediment. These results indicate that the C_{13} chloro-n-alkanes are less

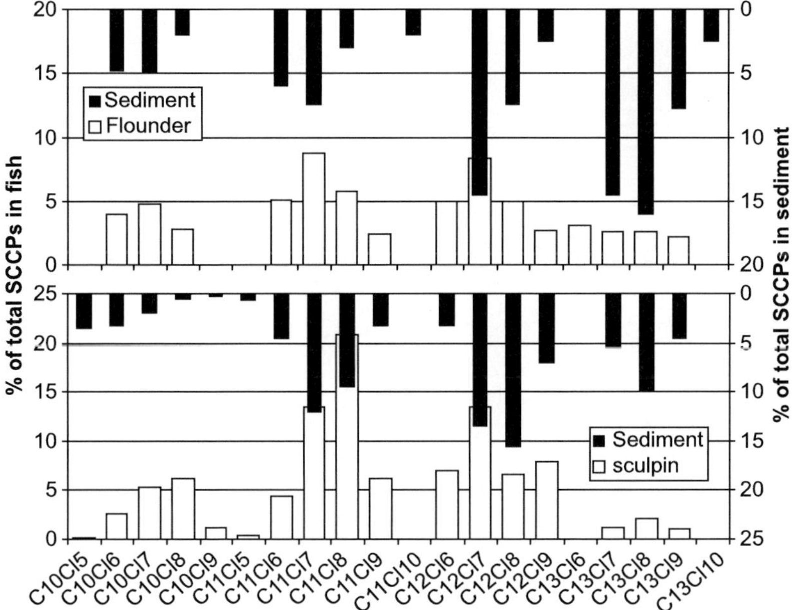

Fig. 6 Comparison of SCCP homologs in sediment (black bars) versus bottom feeding fish (flounder and sculpin) from the same area. Results for flounder from the Baltic Sea are from Reth et al. [54] and for sculpin from Lake Ontario from Houde et al. [59]. Sediment results from the Baltic Sea are from Hüttig and Oehme [35] and from Lake Ontario from Marvin et al. [39]

bioavailable than the shorter chain length SCCPs. There is also an indication that the less chlorinated chain lengths, e.g., $C_{10}Cl_5$, $C_{11}Cl_5$ are underrepresented, possibly due to biotransformation. This transformation might occur in sediment dwelling invertebrates preyed upon by the fish as well as in the fish. Fisk et al. [62] showed that C_{12}–C_{16} chloro-n-alkanes were biotransformed in aerobic sediments and by oligochaetes, and that the susceptibility to degradation in sediments decreased with increasing chlorine content.

Houde et al. [59] studied the accumulation of SCCPs and MCCPs in the Lake Ontario and Lake Michigan food webs. They were able to calculate trophic magnification factors (TMFs) for 31 SCCP–MCCP homologs that were detectable in invertebrates, forage fish, and lake trout. TMFs were evaluated based on the regressions between log concentrations of CP (on a lipid weight basis) and trophic levels of the organisms (derived from $\delta^{15}N$ for the individual organisms analyzed; [63]). A TMF >1 indicates biomagnification is occurring in aquatic food webs. TMFs ranged between 0.47 and 1.5 for SCCPs (17 homologs) and 0.06 and 0.36 for MCCPs (14 homologs) in the Lake Ontario food web. In Lake Michigan, TMFs for SCCPs ranged from 0.41 to 2.4, while results for TMFs could not be evaluated for MCCPs due to the large number of nondetectable values. Statistically significant relationships between trophic level of organisms and log CP concentrations (TMF >1) were observed at both locations for 8 and 9 chlorine substituted decanes,

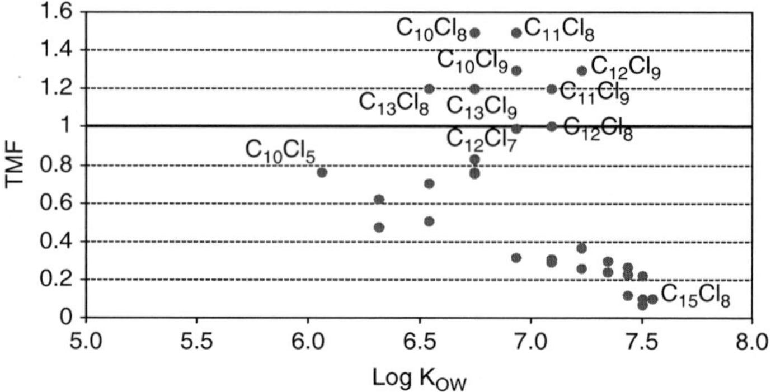

Fig. 7 Relationship of trophic magnification factors for SCCP and MCCP homologs in the Lake Ontario food web with Log K_{OW}

undecanes, and dodecanes. Figure 7, shows the relationship of TMF for the 31 SCCP and MCCP homologs in the Lake Ontario food web with log K_{OW}. No clear relationship with log K_{OW} i.e., with hydrophobicity, was found because the extent of chlorine substitution and the (unknown) congener pattern within each homolog group, are the key characteristics influencing biomagnification to forage fish and lake trout. There was evidence of biotransformation of SCCPs in lake trout based upon lower concentrations than found in forage fish. TMFs calculated without lake trout were generally higher than with them included in the calculation [59].

Temporal trends in fish: Little information exists on temporal trends of CPs in biota with only one study in the peer reviewed literature as of late 2009. Ismail et al. [64] included SCCPs and MCCPs in a study of trends of halogenated flame retardant chemicals in Lake Ontario lake trout. Samples collected approximately every 5 years from 1979 to 2004 were analyzed and thus, only general trends could be assessed. SCCP concentrations were highest in 1998 while MCCP levels in 1993 and 1998 were higher than in other years. Figure 8 compares the time trends of total SCCPs in lake trout and a dated sediment core from the west basin of Lake Ontario. SCCP concentrations in lake trout reported by Houde et al. [59] are included. The maximum SCCP concentration in fish occurs later than observed in sediment.

A number of factors may be responsible for this difference in SCCP maxima. The sediment represents a continuous record of deposition and might be better represented by a series of bars representing individual slices rather than curved lines. The results for lake trout are for individual years and possibly analysis of samples from other years would change the shape of the curve. Ismail et al. [64] noted that food-web processes may influence the interpretation of temporal trends, because the Lake Ontario food web has changed since the 1970s and trophic position of the lake trout as indicated by $\delta^{15}N$ shifted over the years. A shift to more pelagic fish, e.g., alewife (*Alosa pseudoharengus*) and rainbow smelt

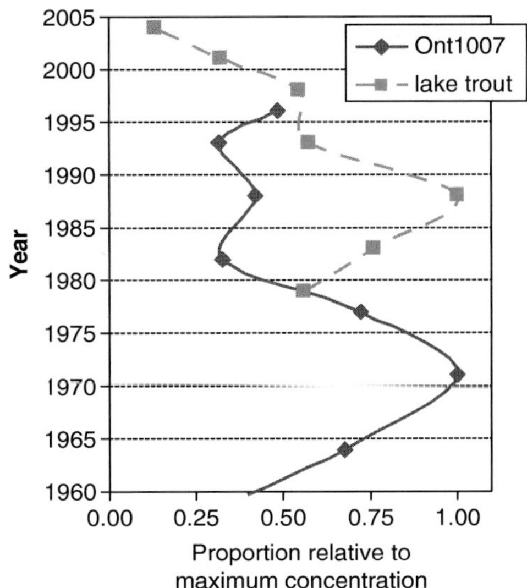

Fig. 8 Comparison of time trends of total SCCPs in lake trout and in a dated sediment core from the west basin of Lake Ontario. Combined results for SCCPs in lake trout from Ismail et al. [64] and Houde et al. [59] and for sediment from Marvin et al. [39]. Results are expressed as a fraction of maximum observed concentration

(*Osmerus mordax*) and away from slimy sculpin might result in lower exposure to SCCPs because of the high concentrations in sculpin. In fact, sculpin populations in Lake Ontario have declined since the early 1990s due to reductions in productivity brought on by nutrient abatement by municipalities on the lake and to reductions in *Diporeia*, an important food of slimy sculpin [65].

4.3 Air, Vegetation, and Soil

Compared to sediments and biota, there is relatively little published information on CPs in air, vegetation, and soil. The available information to the mid-2000s has been thoroughly reviewed in risk assessments [4–6] and in the peer reviewed literature [10, 11].

The only detailed studies of CPs in air have been conducted in the UK. Utilizing high-volume and passive air sampling, Barber et al. [66] demonstrated that SCCPs and MCCPs were major organohalogen contaminants in the UK atmosphere with levels (at their site near Lancaster in northern England) in the same order of magnitude as polyaromatic hydrocarbons and far higher than PCBs or other POPs. Levels of the SCCP in the hi-vol (24 h) samples ranged from <185 to 3,430 pg/m^3 (average 1,130 pg/m^3). MCCP concentrations were higher than SCCPs ranging from <811 to 14,500 pg/m^3 (average 3,040 pg/m^3). Passive air sampling also revealed similar high amounts in other sites throughout England particularly in an urban area where there had been CP production. An earlier study at the same site in northern England, Peters et al. [67] found mean concentrations of SCCPs over a 1 year sampling period (1997–1998) of 320 pg/m^3.

Similar relatively high concentrations of SCCPs were reported by Tomy [68] for extracts of air samples collected in 1990 from the Great Lakes region of Canada $(65–924 \text{ pg/m}^3)$. By contrast much lower levels have been found in the Arctic. Alaee et al. [69] reported average SCCP concentrations of 2 pg/m^3 for the period 1994–1995 at Alert on northern Ellesmere Island, while Borgen et al. [70] reported SCCP concentrations in hi-vol samples from the Mt. Zeppelin (Svalbard) ranging from 9 to 57 pg/m^3. Unfortunately none of the atmospheric studies of SCCP–MCCPs have reported individual homolog concentrations or proportions. This detailed information would be useful for assessing possible sources and transformations when compared to the pattern of CP homologs in commercial products.

CPs have relatively high estimated log K_{OA} values ranging from 9 to 11 for SCCPs and 11–15 for MCCPs [6] as shown in Table 3 for hexachlorodecanes. Coupled with the high concentrations in air, the high log K_{OA}'s imply that vegetation and soils should be accumulating SCCPs and MCCPs by absorption and also by atmospheric particle deposition. Peters et al. [67] found only 1.4–5.1% of total SCCPs concentrations in air on particles when measured by hi-vol sampling but higher proportions would be expected for the MCCPs.

There is very limited information with which to confirm the potential for CPs to accumulate in terrestrial food webs or even for the occurrence in vegetation [5]. The most detailed study to date is by Iozza et al. [71] who analyzed SCCP–MCCP–LCCPs in humus (mean depths: 0.8–23.8 cm) and spruce needle samples along seven different altitude profiles in the Alps. CPs were found in all samples at concentrations ranged from 7 to 199 ng/g (dry weight (dw)) in humus and from 26 and 460 ng/g dw in needle samples, respectively. In humus, CP concentrations were highest at low altitudes (700–900 m above sea level (a.s.l.)). A second maximum was observed between 1,300 and 1,500 m a.s.l., though no trend was found for needles. Concentrations of individual CP groups or homologs were not reported for humus or needles. The updated EU assessment of SCCPs [5] reported SCCPs in three samples of moss from Norway. The samples were taken from remote, forested locations, and had concentrations of SCCPs ranging from 3–100 ng/g.

The high K_{OA} values coupled with relatively high log K_{OW} values put the SCCPs and MCCPs in the class of chemicals that could bioaccumulate in air-breathers in mammalian terrestrial food webs, i.e., from plants to herbivores to top predators such as humans or wolves [72, 73]. The Environmental Bioaccumulation Potential (EBAP) of selected CPs assuming humans exposed via a marine and agricultural diet is shown in Fig. 9. Most SCCPs can be seen to fall into the category of >60% maximum EBAP. The EBAP values generated by the model vary depending on human dietary habits, the structure of the key food chains, and the properties of the environment in which the chemical is distributed.

The available results for air and remote vegetation/humus suggest the ubiquitous nature of CP contamination of the terrestrial environments in Europe. An early report on CPs in wildlife in Sweden by Jansson et al. [74] reported relatively high concentrations in single pooled samples of mammalian herbivores, rabbit

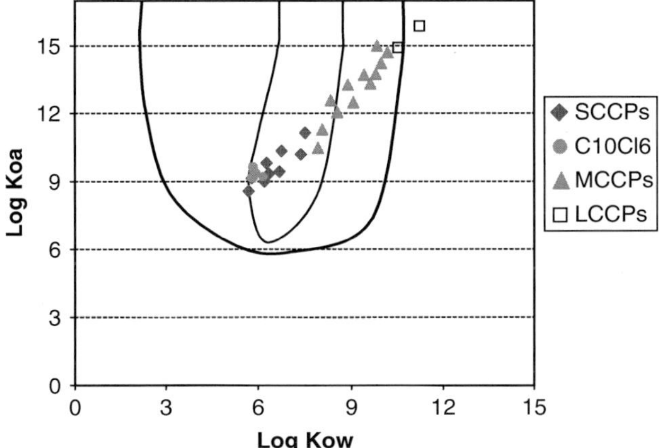

Fig. 9 Assessment of the Environmental Bioaccumulation Potential (EBAP) of selected CPs assuming humans exposed via a marine and agricultural diet based on graphical results presented in Czub and McLachlan [72]. The log K_{OA} values are calculated from log K_{OW} and log K_{AW} values in Tables 2 and 3. The hexachlorodecanes discussed in Table 3 are shown separately. The outer (bold) isoline defines EBAP of >10% maximum EBAP and the inner gray area >60% of maximum EBAP

(*Oryctolagus cuniculus*), and moose (*Alces alces*), 2,200 and 4,400 ng/g lipid wt, respectively. The CP concentration in moose was 30-fold higher than levels in Baltic ringed seal blubber. The concentrations of individual CP chain length or homolog groups were not reported. The results imply a significant accumulation of CPs in the terrestrial food web, as predicted by EBAP estimates, but as of 2009 these results have not been followed up.

5   Conclusions

This review has focused on interpretation of the published results on environmental distribution and spatial-temporal trends of CPs along with general predictions of environmental fate and distribution based on the best estimates of physical properties and biodegradation rates. There are many knowledge gaps regarding the environmental levels and fate of CPs. These can generally be traced to challenges in the analysis of CPs which have clearly deterred many investigators. The development of methods using low resolution MS/MS [75, 76] may encourage more investigators in the field.

The recent risk assessment of SCCPs [5] has concluded that there is a need for further information and/or testing for soils, marine sediments due to uncertainties in the information available for risk assessment. In the case of MCCPs the updated risk assessment [15] requests more environmental measurements, particularly for biota, and notes that the methods used in the past did not unambiguously determine MCCPs. A recent detailed assessment of LCCPs has also noted the lack of measurements and the need for more information and/or testing for soils and marine sediments due to uncertainties [15]. Clearly the lack of environmental levels and fate information, while not the only knowledge gaps, are a major factor in the uncertainties of risk assessments for these substances.

Lack of congener-specific analysis hampers efforts to fully understand fate in sediments as well as food web bioaccumulation when compared to other organohalogen compounds with similar properties such as PCBs and chlorinated pesticides. Analysis at the congener level does not appear to be an achievable goal in the near term due to the complexity of the CP products. However, future investigations, whether by MS/MS or ECNI high resolution MS, should report chain length and homolog information in order to help verify model predictions and assess possible sources, transformations, and fate. The reporting of total SCCPs or total CPs does not permit much assessment due to their wide range of properties and potential sources. If the tri- and tetra chlorinated n-alkanes could be measured, for example, it might be possible to assess environmental half-lives in sediments using the same approach that has been applied to DDT/DDD and hexa- and heptachloro- toxaphene congeners.

The geographical coverage of environmental levels and trends information is limited mainly to Canada/Great Lakes, Canadian and Norwegian Arctic, and Western Europe, with the exception of one detailed study from Japan. Given the large production of CPs in China and possibly in India, there is an urgent need for more studies on environmental levels of CPs in Asian countries. Chinese CP production capacity was 300,000 t in 2003 [77] and reached 600,000 t in 2006 [78]. Actual CP output in China in 2003 was estimated at 150,000 t [77], which exceeds the combined European and North American production of about 100,000 t/year in the late 1990s [6]. From the European and North American experience significant environmental releases can be expected in Asian countries. These studies will need to take into account possible differences between the Asian, North American, and European CP products due to different paraffin feed stocks and production processes [77].

Sediments and biota remain priorities for future CP monitoring according to risk assessments. Analysis of dated sediment cores from Europe and from Asia would help assess the current and past deposition in aquatic environments. Temporal trends of CPs in biota need to be studied and there are a number of wildlife and human tissue banks that could supply suitable samples. While there has been much focus on levels in biota in aquatic environments, more work is needed on levels in terrestrial biota to follow up the early work that demonstrated high concentrations in herbivores. This should include more measurements of human tissue samples (blood, mother's milk) given the potential for human exposure via house dust, vegetation, and meat of herbivorous animals.

To fully understand the environmental distribution of CPs, further information is needed on levels in air, particularly for MCCP and LCCPs. There is growing recognition that flame retardants and plasticizers can be released from products during their life cycle but rates of release are uncertain. Measurements of indoor air and dust would help address these uncertainties.

References

1. Campbell I, McConnell G (1980) Environ Sci Technol 14:1209
2. US EPA (1993) RM2 exit briefing on chlorinated paraffins and olefins. Environmental Protection Agency, Office of Pollution and Prevention of Toxics, Washington, D.C., p 42
3. European Commission (1999) European Union Risk Assessment Report: Alkanes, C10-13, chloro-, Vol. 4. European Commission, European Chemicals Bureau, Ispra, Italy, p 166
4. European Commission (2005) European Union Risk Assessment Report: alkanes, C14-17, chloro (MCCP), Vol. 58. European Commission, European Chemicals Bureau, Ispra, Italy, p 258
5. European Commission (2008) European Union Risk Assessment Report: alkanes, C10-13, chloro (SCCP), Updated version 2008. Vol. 81. European Commission, European Chemicals Bureau, Ispra, Italy, p 148
6. Environment Canada (2004) Follow-up report on a PSL1 substance for which there was insufficient information to conclude whether the substance constitutes a danger to the environment. Chlorinated Paraffins. Environment Canada, Existing Substances Division, Ottawa, ON, p 170
7. UNEP (2009) Revised draft risk profile: short-chained chlorinated paraffins. UNEP/POPS/POPRC.5/INF/18. United Nations Environment Program, Stockholm Convention on Persistent Organic Pollutants Geneva, CH, p 32
8. UNEP (2009) United Nations Environment Program, Stockholm Convention on Persistent Organic Pollutants, Geneva, CH
9. Tsunemi K, Tokai A (2007) J Risk Res 10:747
10. Bayen S, Obbard JP, Thomas GO (2006) Environ Int 32:915
11. Feo ML, Eljarrat E, Barceló D (2009) Trends Analyt Chem 28:778
12. IPCS (1996) Chlorinated paraffins. Environmental Health Criteria 181. International Programme on Chemical Safety, World Health Organization, Geneva, CH
13. Tomy GT, Fisk AT, Westmore JB, Muir DCG (1998) Rev Environ Contam Toxicol 158:53
14. Muir D, Stern G, Tomy G (2000) In: Paasivirta J (ed) The handbook of environmental chemistry, vol. 3 Part K New types of persistent halogenated compounds. Springer, Berlin
15. UK Environment Agency (2009) Environmental risk assessment: longchain chlorinated paraffins. Environment Agency, Wallingford, UK
16. US EPA (2005) Toxics Release Inventory (TRI) Program. US Environmental Protection Agency, TRI Program Division, Washington, DC
18. Nicholls CR, Allchin CR, Law RJ (2001) Environ Pollut 114:415
19. Bergman A, Hagman A, Jacobsson S (1984) Chemosphere 13:237
20. Tysklind M, Soderstrom G, Rappe C, Hagerstedt LE, Burstrom E (1989) Chemosphere 19:705
21. Eurochlor (2008) About Chlorinated paraffins http://www.eurochlor.org/aboutparaffins. Eurochlor, Chlorinated Paraffins Sector Group, Brussels, BE
22. Mackay D, Di Guardo A, Paterson S, Cowan CE (1996) Environ Toxicol Chem 15:1627
23. European Commission (2003) Technical Guidance Document on Risk Assessment. Joint Research Centre, Institute for Health and Consumer Protection, European Chemicals Bureau, Ispra, IT
24. US EPA (2008) EPISuite. US Environmental Protection Agency, Office of Pollution Prevention and Toxics, Washington, DC
25. Thompson RS, Noble H (2007) Short-chain chlorinated paraffins (C10-13, 65% chlori.nated): Aerobic and anaerobic transformation in marine and freshwater sediment systems. Draft Report No BL8405/B. Limited. Brixham Environmental Laboratory, AstraZeneca, UK

26. OECD (2006) OECD Pov & LRTP Screening Tool 2.0. Software and Manual. Organization for Economic Cooperation and Development (OECD), www.oecd.org/env/riskassessment, Paris, France
27. Hilal SH, Karickhoff SW, Carreira LA (2004) QSAR Comb Sci 23:709
28. Sijm DTHM, Sinnige TL (1995) Chemosphere 31:4427
29. Drouillard KG, Tomy GT, Muir DCG, Friesen KJ (1998) Environ Toxicol Chem 17:1252
30. Boethling RS, Howard PH, Meylan WM, Stiteler W, Beauman J, Tirado N (1994) Environ Sci Technol 28:459
31. Klasmeier J, Matthies M, MacLeod M, Fenner K, Scheringer M, Stroebe M, Le Gall AC, McKone T, Van de Meent D, Wania F (2006) Environ Sci Technol 40:53
32. Shojania S (1999) Chemosphere 38:2125
33. Beaume F, Coelhan M, Parlar H (2006) Anal Chim Acta 565:89
34. Tomy GT, Stern GA, Muir DCG, Fisk AT, Cymbalisty CD, Westmore JB (1997) Anal Chem 69:2762
35. Hüttig J, Oehme M (2006) Chemosphere 64:1573
36. El-Morsi TM, Emara MM, Abd El Bary HMH, Abd-El-Aziz AS, Friesen KJ (2002) Chemosphere 47:343
37. Castells P, Parera J, Santos FJ, Galceran MT (2008) Chemosphere 70:1552
38. Hüttig J, Oehme M (2005) Arch Environ Contam Toxicol 49:449
39. Marvin CH, Painter S, Tomy GT, Stern GA, Braekevelt E, Muir DCG (2003) Environ Sci Technol 37:4561
40. Iino F, Takasuga T, Senthilkumar K, Nakamura N, Nakanishi J (2005) Environ Sci Technol 39:859
41. Tomy GT, Stern GA, Lockhart WL, Muir DCG (1999) Environ Sci Technol 33:2858
42. Stern GA, Braekevelt E, Helm PA, Bidleman TF, Outridge PM, Lockhart WL, McNeeley R, Rosenberg B, Ikonomou MG, Hamilton P, Tomy GT, Wilkinson P (2005) Sci Total Environ 342:223
43. Přibylová P, Klánová J, Holoubek I (2006) Environ Pollut 144:248
44. Stern G, Evans M (2003) B.3 Persistent organic pollutants in marine and lake sediments. In Bidleman T, Macdonald R, Stow J (eds) Sources, Occurrence, Trends and Pathways in The Physical Environment. Indian and Northern Affairs Canada, Ottawa, ON, pp 100–116
45. Moore S, Vromet L, Rondeau B (2004) Chemosphere 54:453
46. Iozza S, Muller CE, Schmid P, Bogdal C, Oehme M (2008) Environ Sci Technol 42:1045
47. Muir D, Bennie D, Teixeira C, Fisk A, Tomy G, Stern G, Whittle M (2001) Short Chain Chlorinated Paraffins: Are They Persistent and Bioaccumulative? ACS Symposium Series 773:184–202
48. Rawn DFK, Muir DCG, Savoie DA, Rosenberg GB, Lockhart WL, Wilkinson P (2000) J Great Lakes Res 26:3
49. Rodier D (1999) OECD Expert meeting on chloroparaffins and nonylphenol ethoxylates, Paris, France
50. Walker MJ (2000) OECD Expert Meeting on chloroparaffins and nonylphenol ethoxylates, Paris, France
51. Rose NL, Backus S, Karlsson H, Muir DC (2001) Environ Sci Technol 35:1312
52. Wong CS, Pakdeesusuk U, Morrissey JA, Lee CM, Coates JT, Garrison AW, Mabury SA, Marvin CH, Muir DCG (2007) Environ Toxicol Chem 26:254
53. Oliver BG, Charlton MN, Durham RW (1989) Environ Sci Technol 23:200
54. Reth M, Zencak Z, Oehme M (2005) Chemosphere 58:847
55. Reth M, Ciric A, Christensen GN, Heimstad ES, Oehme M (2006) Sci Total Environ 367:252
56. Evenset A, Christensen GN, Skotvold T, Fjeld E, Schlabach M, Wartena E, Gregor D (2004) Sci Total Environ 318:125
57. Evenset A, Carroll J, Christensen GN, Kallenborn R, Gregor D, Gabrielsen GW (2007) Environ Sci Technol 41:1173
58. Dick TA, Gallagher C, Tomy GT (2009) Intern J Environ Waste Manag (in press)

59. Houde M, Muir DCG, Tomy GT, Whittle DM, Teixeira C, Moore S (2008) Environ Sci Technol 42:3893
60. Fisk AT, Cymbalisty CD, Tomy GT, Muir DCG (1998) Aquat Toxicol 43:209
61. Fisk AT, Tomy GT, Cymbalisty CD, Muir DCG (2000) Environ Toxicol Chem 19:1508
62. Fisk AT, Wiens SC, Barrie Webster GR, Bergman Ã, Muir DCG (1998) Environ Toxicol Chem 17:2019
63. Fisk AT, Hobson KA, Norstrom RJ (2001) Environ Sci Technol 35:732
64. Ismail N, Gewurtz SB, Pleskach K, Whittle DM, Helm PA, Marvin CH, Tomy GT (2009) Environ Toxicol Chem 28:910
65. Mills EL, Casselman JM, Dermott R, Fitzsimons JD, Gal G, Holeck KT, Hoyle JA, Johannsson OE, Lantry BF, Makarewicz JC, Millard ES, Munawar IF, Munawar M, O'Gorman R, Owens RW, Rudstam LG, Schaner T, Stewart TJ (2003) Can J Fish Aquat Sci 60:471
66. Barber JL, Sweetman AJ, Thomas GO, Braekevelt E, Stern GA, Jones KC (2005) Environ Sci Technol 39:4407
67. Peters AJ, Tomy GT, Jones KC, Coleman P, Stern GA (2000) Atmos Environ 34:3085
68. Tomy GT (1997) Ph.D. Thesis. Department of Chemistry, University of Manitoba, Winnipeg, MB Canada
69. Alaee M, Muir D, Cannon C, Helm P, Harner T, Bidleman T (2003) In: Bidleman T, Macdonald R, Stow J (eds) Sources, occurrence, trends and pathways in the physical environment. Indian and Northern Affairs Canada, Ottawa, ON
70. Borgen AR, Schlabach M, Gundersen H (2000) Organohalogen Compd 47:272
71. Iozza S, Schmid P, Oehme M, Bassan R, Belis C, Jakobi G, Kirchner M, Schramm KW, Kräuchi N, Moche W, Offenthaler I, Weiss P, Simončič P, Knoth W (2009) Environ Pollut 157:3225–3231
72. Czub G, McLachlan MS (2004) Environ Sci Technol 38:2406
73. Kelly BC, Ikonomou MG, Blair JD, Morin AE, Gobas FAPC (2007) Science 317:236
74. Jansson B, Andersson R, Asplund L, Litzén K, Nylund K, Sellström U, Uvemo U-B, Wahlberg C, Wideqvist U, Odsjö T, Olsson M (1993) Environ Toxicol Chem 12:1163
75. Zencak Z, Oehme M (2006) Trends Anal Chem 25:310
76. Iozza S, Schmid P, Oehme M (2009) Environ Pollut 157:3218–3224
77. Tang E (2004) China Chemical Reporter Dec 6
78. Jiang G (2009) International Symposium on Organohalogens and Persistent Organic Pollutants August 2009. Beijing, China

Overview of the Mammalian and Environmental Toxicity of Chlorinated Paraffins

Tamer El-Sayed Ali and Juliette Legler

Abstract Chlorinated paraffins (CPs) are a large and complex group of polychlorinated *n*-alkanes found ubiquitously in the environment. Recent studies also show the presence of CPs in human samples such as breast milk. Relatively few studies have been performed on the mammalian and environmental toxicity of CPs. Though the acute toxicity of these compounds is generally low, chronic toxicity and sub-lethal effects have been reported, including carcinogenic effects of short-chained CPs (SCCPs). Target organs for toxicity include the liver, kidneys, and the thyroid and parathyroid glands. Toxicity appears to be inversely related to chain length and increases with greater degrees of chlorination. Lowest effect levels (LOEL) reported in rats are 100 mg/kg/day for the SCCPs (C_{10-13}, 58% Cl, 14 day, oral gavage) and 25 mg/kg/day for medium-chained CPs (C_{14-17}, 52% Cl, 90 days, dietary exposure). The most sensitive endpoint for CP mammalian toxicity appears to be developmental toxicity, with a LOEL in rats of 5.7 mg/kg/day. Mammalian risk quotients of merely 2.63 have been reported based on this LOEL, indicating a relatively small margin of safety between exposure and effect concentrations in mammals. Aquatic invertebrates appear to be highly sensitive to CP toxicity (LOEC 1.6 µg/L for mysids, 28 day exposure to SCCPs C_{10-13}, 58% Cl), and there are indications that fish are highly sensitive to CPs following prolonged exposure to low concentrations. In recent years, a number of comprehensive reviews have been written on CP toxicity. This overview provides a synthesis of the existing information on CP toxicity, and argues that, despite the data gaps, it is clear that, especially SSCPs may lead to significant adverse environmental and human health effects.

J. Legler (✉)
Institute for Environmental Studies, VU University Amsterdam, De Boelelaan 1087, Amsterdam, 1081 HV, The Netherlands
e-mail: juliette.legler@ivm.vu.nl

T. El-Sayed Ali
Oceanography Department, Faculty of Science, Alexandria University, Alexandria, Egypt

J. de Boer (ed.), *Chlorinated Paraffins*, Hdb Env Chem (2010) 10: 135–154,
DOI 10.1007/698_2010_56, © Springer-Verlag Berlin Heidelberg 2010,
Published online: 9 March 2010

Keywords Chlorinated paraffins, Toxicity, Rodents, Fish, Invertebrates

Contents

1   Introduction

Chlorinated paraffins (CPs) are a complex mixture of polychlorinated n-alkanes, having carbon chain lengths ranging from 10 to 38 and a chlorine content ranging from about 30–70% (by weight). They are classified according to their carbon chain length into short-chain CPs (SCCPs, C_{10}–C_{13}), medium-chain CPs (MCCPs, C_{14}–C_{17}) and long chain CPs (LCCPs, $C_{>17}$). There are over 2,000 commercial products containing complex mixtures of CPs [1]. Generally, the average percentage CPs is known for a commercial mixture, though other chlorinated compounds may be present in the mixture. For example, a product described as SCCP, 40% chlorine, will, on average, be composed of chlorinated alkanes that are 40% chlorine by weight and contain predominantly chain lengths between 10 and 13 carbons. However, the product may also contain lower and higher chlorinated alkanes as impurities [2]. CPs were introduced in the 1930s, and are still in use nowadays for a wide range of industrial applications worldwide, such as plasticizers and flame retardants. Another major market for CPs is extreme-pressure additives in metal-working fluids to lower the heat and allow faster metal working. Smaller applications for CPs include flame-retardant additives in rubber, paints, adhesives and sealants. In 2003, the production was estimated to be 300 kt per year worldwide [3].

CPs are aliphatic compounds that differ in physical–chemical characteristics from other widely studied aromatic hydrocarbons such as polychlorinated biphenyls (PCBs), brominated flame retardants and dioxins. Unlike the aromatic hydrocarbons for which important receptor-based mechanisms of toxicity have been elucidated, CP toxicity studies have been carried out mainly with complex technical mixtures and little is known of their toxic mechanisms. In addition, these technical mixtures are likely to contain impurities [4], making the underlying congener-specific mechanism of CPs difficult to decipher. Various detailed reviews have been written on CPs, which present an overview of their toxicity. These include, among others, the "Priority Substances Assessment Report" by the Canadian government [5], an Environmental Health Criteria report by the International Program on Chemical Safety of the World Health Organization [6] and both a European and United Nations Environmental Program risk assessment document regarding short-chained CPs [2, 7]. In addition, an excellent scientific review of the environmental chemistry and toxicology of CPs was written by Tomy and colleagues in 1998 [8]. The objective of this present review is to synthesize the data on mammalian and ecotoxicity provided by these comprehensive reviews. In addition, we will focus on recent data published in the last 10 years, in particular on advances in the environmental toxicology of these complex substances. The limited information on epidemiological studies of human exposure and effects of CPs are summarized in the discussion below.

2 Mammalian Toxicity

The acute toxicity of CPs is low, with oral LD_{50} values in mice and rats exceeding 4 g/kg/bodyweight (bw) (reviewed in [5, 6]). Inhalation or dermal exposure also shows low acute toxicity. For example, an LC_{50} could not be established in the one reported study of inhalation exposure of rats to a concentration of 3,300 mg/m^3 of Chlorowax 500°C (C_{12};59% Cl, CP-SH) for 1 h [9]. Dermal exposure to this SSCP resulted in an LD_{50} exceeding 13 g/kg bw [9].

Toxicity studies of CPs examining non-lethal effects indicate that target organs include the liver, kidneys, and the thyroid and parathyroid glands and that toxicity is inversely related to chain length and possibly increases with greater degrees of chlorination. Toxicity studies with mammals (mainly rodents) are summarized in Table 1. For the SCCPs, the lowest "low effect level" (LOEL) reported is 100 mg/kg/day in rats, for C_{10-13}, 58% Cl (14 day, oral gavage) based on enlarged livers and hepatocellular hypertrophy [12]. In mice, the lowest reported LOEL is 125 mg/kg/day for C_{12}, 58% Cl (2 year, oral gavage) based on adenomas in liver, kidneys, thyroid and other tissues [10, 11]. In a recent study by Hallgren and Darnerud [15], Sprague-Dawley rats exposed by oral gavage to the low level of 6.8 mg/kg/day Witaclor 171P (C_{10-13}; 71% Cl) for 14 days showed no effects on liver or thyroid weights, or on thyroid hormone levels or related biotransformation enzyme induction. However, when administered together with an isomolar concentration of the flame retardant bromodiphenyl ether 47 (BDE 47) (6.0 mg/kg/day) or the PCB mixture Aroclor 1254 (4.0 mg/kg/day), these mixtures significantly influenced thyroid hormone levels and induction of the microsomal phase I enzymes ethoxy- and methoxy-resorufin dealkylases (EROD and MROD). Interaction analysis of the data from the Witaclor-BDE 47 mixture exposure indicated synergistic effects on EROD induction and free thyroxin (T4) levels, which were decreased to

Table 1 Mammalian toxicity of chlorinated parrafins

Species	CP mix tested	Exposure duration/route	Effect	Reference
SCCPs				
F344/N rats, B6C3F1 mice (both sexes)	C_{12}; 60% Cl	16 days gavage in corn oil	LOEL hepatomegaly: 469 mg/kg/day (rats); 938 mg/kg/day (mice); enlarged livers (mice), decreased body weights (rats), and diarrhoea (both species)	[10, 11]
F344 rats	C_{10-13}; 58% Cl	14 days gavage in corn oil	LOEL: 100 mg/kg/day enlarged livers and hepatocellular hypertrophy	[12]
B6C3F1 mice F344/N rats	C_{12}; 60% Cl	13 weeks gavage in corn oil	LOEL: 250 mg/kg/day (mice); 125 mg/kg/day (rats); hepatocellular hypertrophy	[10, 11]
F344 rats	C_{10-13}; 58% Cl	90 days gavage in corn oil or diet	NOEL: 10 mg/kg/day increases in liver and kidney weights, increases in the incidence of hepatocellular hypertrophy, increases in thyroidparathyroid weights, hypertrophy and hyperplasia of the thyroid; high incidences of trace-to-mild chronic nephritis in kidneys of male rats and increased pigmentation of the renal tubules in female rats	[1]
F344/N rats B6C3F1 mice	C_{12}; 58% Cl	2 years gavage in corn oil	LOAEL = 312 mg/kg/day (rats); 125 mg/kg/day (mice) hepatocellular neoplasms and adenomas or adenocarcinomas of the liver; mononuclear cell leukemia; adenomas or hyperplasia of the renal tubular cells in exposed male rats; follicular cell adenomas or carcinomas of the thyroid in exposed female rats and female mice; alveolar/bronchiolar adenomas or carcinomas in male mice	[10, 11]
Wistar rats, male	Paroil 170-HV (C_{11}; 70% Cl)	Intraperitoneal, single dose enzyme activity measured after 24 h, 7 or 21 days	52 mg/kg Paroil: intestinal activities of aryl hydrocarbon hydroxylase (increase), UDP-glucuronosyltransferase (decrease) and epoxide hydrolase (increased)	[13]

F344 rats	Chlorowax 500°C (C_{12}; 58% Cl)	90 days gavage by corn oil	LOEL 313 mg/kg/day; based on increased relative liver weights, hepatic peroxisomal ß-oxidation and thyroxine-UDPG-glucuronosyl-transferase activity	[14]
Sprague-Dawley rats (female)	Witaclor 171P (C_{10-13}; 71% Cl, CP-SH)	14 days gavage by corn oil	6.8 mg/kg/day: no effects liver or thyroid weights, thyroid hormone levels	[15]
MCCPs				
F344 rats	C_{14-17}, 52% Cl	14 days diet	LOEL 177 mg/kg/bw males: increased cytochrome P-450 values; LOEL 562 mg/kg/bw females: increased liver weight and hepatocellular hypertrophy	[1]
F344 rats	C_{14-17}, 52% Cl	90 days gavage in corn oil or diet	LOEL 100 mg/kg/day: increased liver and kidney weights; >10 mg/kg/day: increased liver, kidney and thyroid–parathyroid weights, incidence of hepatocellular hypertrophy; hypertrophy and hyperplasia of the thyroid; high incidences of trace-to-mild chronic nephritis in kidneys of male rats; increased pigmentation of the renal tubules in female rats	[1]
Wistar rats	C_{14-17}, 52% Cl	90 days diet	LOEL: 25 mg/kg/day: increased liver weight, proliferation of the smooth endoplasmic reticulum in hepatic cells	[16]
Beagle dogs	C_{14-17}, 52% Cl	90 days diet	LOEL: 30 mg/kg/day: proliferation of the smooth endoplasmic reticulum in hepatic cells	[16]
Alpk:APFSD rats; Alpk:APFCD-1 mice, All males	Chlorowax 500°C (C_{10-13}; 58%Cl), Cereclor 56 L (C_{0-13}; 56% Cl) Chlorparaf-fin 40G C_{14-17}; 40% Cl)	14 days, gavage by corn oil	LOEL Chlorowax 500°C 100 mg/kg/day, LOEL Cereclor 56 L 50 mg/kg/day, LOEL Chlorparaffin 100 mg/kg/day: increased liver weight; 1,000 mg/kg/day: thyroid function: reduction in plasma T4 levels (both free and total), increase in plasma TSH levels, no effect plasma T_3; increase (twofold) in the rate of glucuronidation of T4 by hepatic microsomal UDP glucuronosyltransferase activity	[17]

(continued)

Table 1 (continued)

Species	CP mix tested	Exposure duration/route	Effect	Reference
Sprague-Dawley rats, weanlings	C_{14-17}; 52%	90 days, diet	LOEL 36 mg/kg/day males: thyroid histopathology LOEL 43 mg/kg/day females: increased liver weight, decreased hepatic vitamin A levels dose levels, thyroid histopathology, kidney effects	[18]
LCCPs				
F344/N rats B6C3F1 mice	C_{23}; 40% Cl	16 days gavage in corn oil	NOEL highest test doses 3,750 mg/kg/day (rats), 7,500 mg/kg/day (mice); no clinical or gross pathological effects	[11, 19]
F344 rats	C_{20-30}; 43% Cl	14 day gavage	NOEL 3,000 mg/kg/day; no clinical signs or effects on organ weights or in tissues examined microscopically, LOEL 3,750 mg/kg/bw males: mild nephrosis; LOEL 100 mg/kg/day females: liver effects	[1, 20, 21]
F344 rats	C_{22-26}; 70% Cl	14 day diet	NOEL: 1,715 mg/kg/day; no clinical signs or effects on organ weights or in tissues examined microscopically LOEL: 3,750 mg/kg/day; hepatocellular hypertrophy and cytoplasmic fat vacuolation in liver and increases in serum hepatic enzymes	[1, 20, 21]
F344/N rats B6C3F1 mice	C_{23}; 43% Cl	13 weeks gavage in corn oil	LOEL rats: 235 mg/kg/day; granulomatous inflammation of liver in all females	[11, 19]
F344/N rats B6C3F1 mice	C_{23}; 43% Cl	2 years gavage in corn oil	No effects in survival and clinical toxicity at doses up to 3,750 and 5,000 mg/kg/day (rats and mice, respectively). Significant increase in the incidence of malignant lymphomas in male mice LOAEL 100 mg/kg/day (female rats); non-neoplastic lesions in liver, pancreatic and mesenteric lymph nodes	[10, 11]
Wistar rats, male	Chlorez 700 (C_{20}; 70% Cl, 52 mg/kg	Intraperitoneal, single dose enzyme activity measured after 24 h, 7 or 21 days	100 mg/kg/bw: increased kidney activity of aryl hydrocarbon hydroxylase	[13]

levels of 51% of the control. Though a mechanism for these synergistic effects is not given, the authors suggest that CPs could potentially modulate other important microsomal enzyme activities, which may underlie the decrease in plasma T4 [15]. This study has important implications for the risk assessment of CPs, as it indicates that their toxicity may be modulated by the presence of other organohalogens in mixtures.

For the MCCPs, rodent toxicity studies have revealed similar or even lower LOEL levels than SCCPs (Table 1). The lowest reported LOEL is 25 mg/kg/day in Wistar rats given C_{14-17}, 52% Cl in the diet for 90 days [16]. Effects observed at this LOEL were similar to SCCPs, that is increased liver weight and proliferation of the smooth endoplasmic reticulum in hepatic cells. For the LCCPs, reported effect concentrations are generally higher than for SC- and MCCPs, though LOELs of 100 mg/kg/day have also been reported (Table 1). Exposure of female Wistar rats to C_{20-30}, 43% Cl (14 day gavage) showed effects on liver weights at this dose [1, 20, 21].

2.1 Carcinogenicity

In an evaluation by the International Agency for Research on Cancer (IARC), it was concluded that SCCPs of average carbon chain length C_{12} and average degree of chlorination of approximately 60% are possibly carcinogenic to humans [22]. This conclusion was based on effects reported in experimental animals, such as those reported by the National Toxicology Program (NTP) study of F344/N rats and B6C3F1 mice exposed for 2 years (Table 1; [11, 12]). However, IARC also concluded that there was (too) limited evidence for the carcinogenicity of a commercial CP product of average carbon chain length C_{23} and average degree of chlorination of 43% in experimental animals [22]. The NTP study of this LCCP concluded that there was no evidence of carcinogenicity for male F344/N rats, but equivocal evidence of carcinogenicity for female F344/N rats and female B6C3F1 mice (Table 1; [11, 12]).

2.2 Genotoxicity

The mechanisms underlying CP carcinogenicity are unclear. Though only limited studies have been carried out, mutagenic effects appear to be absent for SSCPs, MCCPs or LCCPs in bacterial assays in vitro with or without metabolic activation [4, 10]. In mammalian cells, the SSCP C_{12}; 60% Cl has been found to be mutagenic in L5178Y mouse lymphoma cells at concentrations of 48 and 60 μg/mL in the absence of metabolic activation [23]. Chlorowax 500°C (C_{23}; 43% Cl) induced chromosome aberrations in the absence of S9 mix at a concentration of

5,000 µg/mL and sister chromatid exchange (LOEC 5 µg/mL) with and without S9 in Chinese hamster ovary in vitro [24]. Another in vitro study has shown that SC- and MCCPs, but not LCCPs, are potent inhibitors of gap junction intercellular communication, suggesting that these compounds may act as tumour promoters [25]. In this study, six commercial CPs of different carbon chains length and chlorine content (Cereclor 50LV (C50LV, C_{10-13}; 56% Cl), Hüls 60 (H60, (C_{10-13}; 60% Cl), Cereclor S45 (CS45, C_{14-17}, 45% Cl), Cereclor S52 (CS52, C_{14-17}, 52% Cl), Cereclor 42 ($C_{>20}$, 42% Cl) and Cereclor 48 ($C_{>20}$, 48% Cl) on cell communication were studied in IAR 20 rat liver epithelial cells. At non-cytotoxic concentrations, C50LV, H60, CS45 and CS52 completely inhibited the cell communication within 1 h [25].

In an in vivo study with rats, SCCPs, MCCPs and LCCPs did not induce dominant lethal mutations or increase the frequency of chromosomal aberrations in bone marrow cells [1]. Another in vivo study with SCCPs showed that when C_{12}; 60% Cl was administrated in corn oil by gavage to Alpk:AP male rats, no effects on unscheduled DNA synthesis in hepatocytes could be detected after exposure for 2 or 12 h [26]. However, a moderate dose- and time-related induction of cell proliferation, measured as S-phase cells in hepatocytes was detected in animals exposed to 1,000 and 2,000 mg/kg for 12 h.

Lack of chromosomal aberrations has been shown for CPs studied in mammalian cells collected from animal experiments (reviewed in [6]). Sexually mature male Fischer-344 rats administered the SCCP C_{10-12}; 58% Cl by gavage once daily for 5 days at concentrations up to 750 mg/kg/bw/day, did not show an increased mortality or frequency of chromosome or chromatid abnormalities in bone marrow cells [27]. Similarly, no chromosomal aberrations were found for an MCCP (C_{14-17}; 52% Cl) in the same experimental design at doses up to 5,000 mg/kg/day [28] or in two LCCP formulations (one with 70% chlorination, the other C_{22-26}; 43% Cl) at doses up to 5,000 mg/kg/day [29, 30]. MRI mice given single doses of 50 and 5,000 mg/kg bw Chlorowax 500°C (C_{10-13}; 58% Cl) by gavage showed no effects in polychromatic cells with micronuclei or in the ratio of polychromatic erythrocytes to normocytes [31]. Similar lack of effects in the mouse bone marrow micronucleus assay have been reported for the MCCP mixtures Solvocaffaro C1642 (C_{14-17}; 42% Cl) and Melex DC 029 (C_{14-17}; 45% Cl). Studies with a SCCP with 58% chlorination in Charles River CD rats [32] showed no evidence of a mutagenic effect on the spermatogenesis at oral doses of up to 200 mg/kg/day for 5 days.

2.3 Reproductive Toxicity

The most sensitive endpoint for CP mammalian toxicity reported to date is developmental toxicity, as reported in a study with rats exposed to a MCCP (C_{14-17}, 52% Cl) in the diet for 28 days before mating, during mating, and for females, continuously up to postnatal day 21. Pups were also exposed from weaning to 70 days of age [1, 33]. In this investigation, no differences were found in the parental generation, for example, on appearance, fertility, body weight gain, food consumption or reproductive performance. However, effects on the development of offspring were found at all doses, such as adverse effects on body weight and condition, and haematological parameters (LOEL 5.7 mg/kg/day for the males and 7.2 mg/kg/day for the females). Pup survival was also decreased at doses of 57 mg/kg/day or higher. Effects were more likely attributable to lactational rather than in utero exposure, as preliminary results from a cross-fostering study suggested that mortality in pups exposed via milk was greater than that in pups exposed only in utero [1]. Accordingly, in a reprotoxicity study in which the same MCCP was administered to by gavage to pregnant Charles River rats on days 6–19 of gestation and pregnant Dutch Belted rabbits on days 6–27 of gestation, no effects on the development of the offspring were found. However, embryo- or feto-toxic effects were observed at doses that were greater than those that were toxic to the mothers (lowest NOAEL in mothers was 30 mg/kg/day and in offspring, 100 mg/kg/day) [34, 35].

Reproductive toxicity has also been tested for SSCP (C_{10-13}, 58% Cl), and LCCP (C_{20-30}, 43% Cl and C_{22-26}, 70% Cl) administered by gavage in corn oil to pregnant Charles River rats on days 6–19 of gestation and pregnant Dutch Belted rabbits on days 6–27 of gestation. For the SSCP, in rats, the incidence of adactyly and/or shortened digits was observed in the offspring of rats exposed to a maternally toxic dose (2,000 mg/kg/day) [36]. In rabbits, developmental toxicity was observed in offspring at doses of SSCP that were not toxic to the mothers (30 and 100 mg/kg/day) [37]. In the exposure to LCCP, no teratogenic effects were observed. Embryo- or feto-toxic effects were observed only at doses greater than those that were toxic to the mothers (lowest LOEL in mothers $=$ 100 mg/kg/day in rabbits exposed to the C_{22-26}, 70% Cl CP; lowest NOEL in offspring $=$ 1,000 mg/kg/day in rabbits exposed to the C_{22-26}, 70% Cl CP) [36, 38–40].

In a study of NMRI mice, a single intraperitoneal injection of 100 mg/kg body weight of the MCCP polychlorohexadecane (C_{16}; 70% Cl) on the day the vaginal plug was observed, showed no effects on implantation or embryonic survival [41].

2.4   Neurotoxicity

Very limited data is available on the neurotoxicity of CPs. In one study, decreased motor capacity was observed in adult NMRI male mice administered a single dose 300 mg/kg/bw of SCCP (C_{10-13}, 49% Cl) intraperitoneally [42]. In another study, oral administration of a single dose (1 mg/kg bw) of the MCCP polychlorohexadecane (C_{16}; chlorination degree not specified) to 10-day-old male and female mice showed no effect on muscarinic receptors, though evidence for a presynaptic effect on the cholinergic system was suggested [43]. No data is available on the neurotoxicity of LCCPs.

2.5 *Immunotoxicity*

No data is available on the immunotoxicity of the short-, medium- or long-chain CPs.

3 Environmental Toxicity

The number of studies on the acute or chronic effects of CPs on aquatic and, in particular, terrestrial organisms is quite limited. An overview of the aquatic eco-toxicity of CPs is given in Table 2.

3.1 *Aquatic Organisms*

3.1.1 Algae

Algae show a lower sensitivity to CPs than invertebrates. Both freshwater and marine algae have been exposed a SCCP (C_{10-12} 58% Cl; CP-SH) and effects have been found on growth in 2–10 day exposure experiments (Table 2). The sensitivity of the marine algae exceeded the freshwater species, with an EC_{50} of 43 µg/L in 4 days, compared to 1,310 µg/L [44, 45]. Interpretation of the results in algae is complicated by the loss (50–80%) of SCCP from the water during the experiments, likely due to sorption to algal cells. In addition, the effects observed on the marine algae were transient over the 10-day test period, and may have been caused by a decrease in nutrient levels [45].

3.1.2 Aquatic Invertebrates

Aquatic invertebrates appear to be the most sensitive aquatic species for CP effects (Table 2). The lowest reported "low effect concentration" (LOEC) of 1.6 µg/L has been shown for mysid mortality following chronic (28 day) exposure to SCCPs (C_{10-13}, 58% Cl) [47]. EC50s for the freshwater crustacean Daphnia magna follow-ing exposure from 3 to 21 days to SCCPs (C_{10-13}, 58% Cl) ranged from 24 to 12 µg/ L [46]. After 21 day exposure at 8.9 µg/L (LOEC), though no mortality was seen in adult *Daphnia*, significantly higher mortality was found in the offspring. Mussels *Mytilus edilus* also showed similar sensitivity in chronic exposure experiments, with a LOEC of 9.3 µg/L (reduction of growth) in 84-day flow through experiments. Large differences in sensitivity were found for mussels treated with MCCPs and LCCPs, however. Mussels exposed for 60 days with MCCPs (C_{14-17}; 52% Cl)

Table 2 Aquatic ecotoxicity of chlorinated paraffins[a]

Species	CP mix tested	Exposure duration/ route	Effect	References
Algae				
Marine diatom *Skeletonema costatum*	C_{10-13}; 58% Cl	96-h	EC_{50} 42.3 µg/L; growth	[44]
Freshwater green alga *Selenastrum capricornutum*	C_{10-13}; 58% Cl	10 days	LOEC 570 ug/L: inhibited growth EC_{50} 1,310 µg/L	[45]
Aquatic invertebrates				
Daphnids *Daphnia magna*	C_{10-13}, 58% Cl	96 h flow-through	LC_{50} 18 µg/L	[46]
Daphnids *Daphnia magna*	C_{10-13}; 58% Cl	21 days flow-through	LOEC: 8.9 µg/L; increase in mortality of offspring	[46]
Mysid shrimp *Mysidopsis bahia*	C_{10-13}; 58% Cl	96 h flow-through	LC_{50} 14 µg/L	[47]
Mysid shrimp *Mysidopsis bahia*	C_{10-13}; 58% Cl	28 days flow-through	LOEC: 1.6 µg/L; mortality	[47]
Mussels *Mytilus edilus*	58% Cl	84 day flow-through	LOEC 9.3 µg/L: reduction of growth	[48]
Mussels *Mytilus edilus*	C_{10-12}; 58% Cl	60 days flow-through	LC_{50} 74 µg/L	[49]
Mussels *Mytilus edilus*	C_{14-17}; 52%	60 day flow-through	3,800 µg/L: reduced filtration activity	[50]
Mussels *Mytilus edilus*	43% and 70% Cl	60 day flow-through	LOEC 2,200 µg/L for the 43% chlorinated CP, and 1,330 µg/L for the 70% chlorinated CP: reduced filtration activity	[50]
Midge larvae Chironomus tentans	C_{10-12}; 58% Cl	48 h or 49 days	LOEC 48 h > 162 µg/L LOEC 49 121 µg/L: adult emergence halted	[51]
Fish				
Bleak *Alburnus alburnus*	(a) Witaclor 149 (C_{10-13}; 49% Cl) Witaclor 159 (C_{10-13}; 59% Cl) Witachlor 171P (C_{10-13}; 71% Cl) (b) Witaclor 350 (C_{14-17}; 50% Cl (c) Witaclor 549 (C_{18-26}; 49% Cl)	14 days single concentration (125 µg/L)	(a) 125 µg/L: behavioural effects (b) No observed effect (c) No observed effect	[52, 53]

(*continued*)

Table 2 (continued)

Species	CP mix tested	Exposure duration/ route	Effect	References
Sheepshead minnow *Cyprinodon variegatus*	C_{10-12}; 58% Cl	32 day	LOEC 279.7 µg/L: reduced size of larvae	[54]
Rainbow trout *Oncorhynchus mykiss*	(a) C_{10-12}; 58% Cl (b) C_{14-17}; 52% (c) C_{20-30}; 43%, 70% Cl	60 day flow-through	(a) LC_{50}: 340 µg/L (b) 1,050 µg/L: no effect (c) 3,800 µg/L: no effect	[50]
Flounder *Platichthys flesus* (both sexes)	Witachlor 149 (C_{12}, 49% Cl); Chlorparaffin Hüls 70°C (C_{12};70% Cl)	Dietary exposure on days 1 and 4	After 13 and 27 days: 1,000 mg/kg bw: sublethal effects on the haematology, glucose metabolism and xenobiotic and steroid metabolising enzymes of female fish	[55]
Rainbow trout *Oncorhynchus mykiss*	$C_{10}H_{15.5}C_{6.5}$, $C_{10}H_{15.3}C_{16.7}$, $C_{11}H_{18.4}C_{15.6}$, $C_{12}H_{19.5}C_{16.5}$, $C_{14}H_{24.9}C_{15.1}$ $C_{14}H_{23.3}C_{16.7}$	21 days diet	Loss of equilibrium and dark coloration, histopathological lesions in livers of fish exposed to C_{10} and C_{11}	[56]

[a]Concentrations given are actual, unless indicated otherwise

showed no mortality at the highest test concentration 3,800 µg/L, which also exceeded the maximum water solubility [47]. For the two LCCPs tested (C_{22-26}; 43% Cl and C_{20-30}; 70% Cl), 60 day exposure to the highest test concentrations of 2,180 and 1,330 µg/L caused effects on filtration activity [44, 45]. It should be noted, however, that these concentrations exceeded the maximal solubility of the CPs.

3.1.3   Fish

The acute toxicity of CPs to fish appears to be limited. In a study of CPs of various chain lengths (C_{10-13}; 49%, 63%, and 71% Cl, C_{10-13}; 56% Cl, $C_{11.5}$; 70% Cl, $C_{15.5}$; 40% Cl, C_{14-17}; 50% and 52% Cl, C_{22-26}; 42% Cl, C_{18-26}; 49% Cl), 96 h LC_{50} values in the estuarine bleaker (*Alburnus alburnus*) exceeded the highest test concentrations of 5×10^6 µg/L [57]. In a 96-h study on rainbow trout (*Oncorhynchus mykiss*), no toxic effects or effects on behaviour were observed after exposure of the fish to an emulsion containing a mean concentration of 770,000 µg/L of Cereclor 42 (C_{20-30}; 42% Cl) [58]. Chronic toxicity experiments, however, have indicated that fish may require a longer exposure period for the toxicity of CPs to be exerted. Short-term toxicity tests may underestimate toxicity. For example, in a bioconcentration study by Madeley and Maddock [59], rainbow trout were exposed to concentrations of 3.1 and 14.3 µg/L for 168 days. The fish were removed to fresh water for a depuration period of 105 days. Starting at day 63 of depuration, fish previously exposed to 14.3 µg/L began to exhibit behavioural symptoms. By day 69, all fish exposed to this concentration had died, as well as 50% of those from the group exposed to 3.1 µg/L. Other chronic studies have also revealed that toxicity of SCCPs tends to exceed that of MCCPs and LCCPs (Table 2).

In a recent study, Cooley and coworkers [56] have provided a useful and unique contribution to ecotoxicological data on individual CP congeners of single carbon chain lengths, as most toxicity studies have been performed with technical mixtures. Juvenile rainbow trout (*Oncorhynchus mykiss*) were exposed to six individual SC- and MCCPs ($C_{10}H_{15.5}C_{6.5}$, $C_{10}H_{15.3}C_{16.7}$, $C_{11}H_{18.4}C_{15.6}$, $C_{12}H_{19.5}C_{16.5}$, $C_{14}H_{24.9}C_{15.1}$ and $C_{14}H_{23.3}C_{16.7}$) in the diet at concentrations that have been reported in invertebrates and fish from contaminated sites in the Great Lakes. With the exception of trout exposed to $C_{14}H_{24.9}C_{15.1}$, which had much lower exposure concentrations, trout exposed for 21 days to the individual CPs (whole fish concentrations 0.22–5.5 µg/g) showed effects on behaviour, including altered startle response and feeding, as well as altered liver histopathology. No effects were found on thyroid histology, although trout exposed to $C_{10}H_{15.5}C_{6.5}$ (whole fish concentration 0.84 µg/g) had slightly more active thyroids. These responses observed in this study were indicative of a narcotic toxicological mode-of-action. The authors concluded that individual CP toxicity is inversely related to carbon chain length, as has been observed in CP toxicity studies using mammals [56].

3.1.4 Potential Risk of CPs to Aquatic Organisms

Given the apparent sensitivity of aquatic invertebrates and fish to particular SCCP exposure, the potential risk to these organisms can be calculated. As shown above, 8.9 μg/L was the LOEC for *Daphnia* [46]. UNEP has adopted this LOEC as a "critical toxicity value" for pelagic invertebrates, and calculated a "predicted no effect concentration" (PNEC) of a factor 1,000 lower, i.e. 8.9 ng/L [2]. By comparing the PNEC with the estimated environmental exposure value of 44.8 ng/L, a risk quotient (RQ) of five was reported [2], indicating a relatively small margin of safety between environmental concentrations and effect concentrations of SSCPs in *Daphnia*. Similarly, an RQ was calculated for pelagic fish using data from the Cooley et al. [56] study described above. In this case, an RQ of 3,329 was calculated, indicating a relatively large margin of safety for fish [2].

3.2 Terrestrial Organisms and Birds

For the terrestrial environment, very few studies are available on CP toxicity. Recently, two studies on soil-dwelling organisms have been published. The effects of a technical SCCP formulation characterized as C_{12}; 64% Cl containing C_{10} 6%, C_{11} 37%, C_{12} 32% and C_{13} 25% on invertebrates (*Eisenia fetida, Folsomia candida, Enchytraeus albidus, Enchytraeus crypticus, Caenorhabditis elegans*) and substrate-induced respiration of indigenous microorganisms were studied [60]. Soil organisms were exposed to spiked soil for various time periods, and acute and chronic effects were observed (Table 3). The springtail *F. candida* was the most sensitive organism with LC_{50} and EC_{50} (reproduction) values of 5,733 and 1,230 mg/kg, respectively. These researchers estimated a PNEC for soil of 5.28 mg/kg, which slightly exceeds the estimated exposure value of 3.2 mg/kg reported by UNEP [2], indicating a potential risk to soil dwelling organisms. In another study, Sochová et al. [61] examined the effect of this same SCCP on the nematode *Caenorhabditis elegans* in various media: water, soil and agar (Table 3). Toxicity was more pronounced following 48 h exposure compared to 24 h, and was one order of magnitude lower than LC_{50} values reported for other soil invertebrates from the Bezchlebova study [60].

In birds, based on the limited studies available, both the acute and chronic toxicity of CPs appears to be low. In a one-generation reproductive study in which mallard ducks (*Anas platyrynchos*) were fed 28, 166 and 1,000 mg/kg-diet of a SCCP (C_{10-13}), a slight decrease in eggshell thickness and 14-day embryo viability was found at the highest dose only [1]. In a study of the MCCP Cereclor S52 (C_{14-17}; 52% Cl), the acute oral LD_{50} values were >24,606 mg/kg for ring-necked pheasants (*Phasianus colchicus*) and >10,280 mg/kg for mallard ducks. The acute dietary LC_{50} for the latter species was >24,063 mg/kg-diet [58]. The toxicity of Cereclor 42 (C_{22-26}; 42% Cl, CP-LL), Cereclor 50LV C_{10-13}; 49% Cl, CP-SL) and Cereclor 70 L (C_{10-13}; 70% Cl, CP-SH) in chick embryos was studied

Table 3 Terrestrial ecotoxicity of chlorinated paraffins

Species	CP tested	Exposure duration/route	Effect	References
Folsomia candida	C_{12}; 68% Cl[a]	28 day soil	LOEC Mortality: 2,500 mg/kg Repro: 1,250 mg/kg	[60]
Eisenia fetida	C_{12}; 68% Cl	28 day soil	Mortality: >10,000 mg/kg Repro: 3,200 mg/kg	[60]
Enchytraeus albidus	C_{12}; 68% Cl	42 day soil	Mortality: >10,000 mg/kg Repro: 6,000 mg/kg	[60]
Enchytraeus crypticus	C_{12}; 68% Cl	28 day soil	Mortality: 10,000 mg/kg Repro: 10,000 mg/kg	[60]
Caenorhabditis elegans	C_{12}; 68% Cl	48 h soil	Mortality: 3,000 mg/kg Repro: 10,000 mg/kg	[60]
Caenorhabditis elegans	C_{12}; 68% Cl	24 and 48 h soil	LC_{50} 24 h: 5,450 mg/kg 48 h: 8,833 mg/kg	[61]
Caenorhabditis elegans	C_{12}; 68% Cl	24 and 48 h water	LC_{50} 24 h: > 0.5 mg/L 48 h: 0.5 mg/L	[61]
Caenorhabditis elegans	C_{12}; 68% Cl	24 and 48 h agar	LC_{50} 24 h: 2,372 mg/L 48 h: 869 mg/L	[61]

[a]This technical mixture included all short-chain paraffin fractions ($C_{10–13}$) with a composition of C_{10} 6%, C_{11} 37%, C_{12} 32% and C_{13} 25%

by Brunström [62]. The CPs were injected into the yolks of eggs incubated for 4 days at concentrations of 100 and 200 mg/kg egg. None of the three mixtures affected the hatching rate, incubation time, hatching weight, weight gain after hatching or the liver weights of the chicks. In an extension of this study, eggs were injected with 300 mg/kg egg weight of the same Cereclors [63]. Effects were observed after treatment with the most highly chlorinated SCCP, $C_{10–13}$; 70% Cl, including increased liver weights and microsomal concentration of cytochrome P450. In a recent study by Ueberschär and Matthes [64], broiler chickens exposed to 100 mg/kg SCCP (C_{10}–C_{13}) through the feed for 31 days showed no adverse effects on mortality, organ weights or growth rates. Similarly, in a study by the same authors in which laying hens were exposed to SCCP (C_{10}–C_{13}; 60% Cl) from 24 to 32 weeks of age in increasing concentrations of up to 100 mg/kg feed, no adverse affects were found on health, relative organ weights or performance (laying intensity, egg weight, feed consumption) [65].

4 Toxicity in Humans and Potential Risk

The direct effects of CPs in humans are limited to poorly documented clinical studies from the 1970s of the potential to induce irritation or sensitization of the skin following dermal application (reviewed in [5]). Based on the mammalian toxicity studies outlined above, the lowest effect level reported for CP is 5.7 mg/

kg/day for reproductive toxicity of the MCCP (C_{14-17}; 52% Cl). Based on this LOEL, the "no effect level" (NOEL) can be assumed to be about 1 mg/kg/day. UNEP has also reported 1 mg/kg/day as a PNEC for SSCPs, and has derived a RQ from this PNEC, based on estimated exposure values from contaminated fish [2]. The RQ for mammals was merely 2.63, which indicates that there is a relatively small margin of safety between exposure and effect concentrations in mammals. Environment Canada has also reported a tolerable daily intake (TDI) based on this LOEL of 5.7 mg/kg/day, which when corrected for a safety factor of 1,000, was reported to be 6 μg/kg/day [5].

Importantly, exposure assessment studies have been carried out in recent years have provided evidence of human exposure to these substances and accumulation in humans and foodstuffs (reviewed in [2, 6]). A recent study has found SCCPs (C_{10-13}) to be present in human breast milk samples from the United Kingdom, at concentrations ranging from 49 and 820 ng/g fat (median 180 ng/g fat) [66]. MCCPs (C_{14-17}) were detected at levels of 6.2–320 ng/g fat (median 21 ng/g fat). Assuming a median CP breast milk concentration of 180 + 21 = 201 ng/g fat, and a weighted mean lipid intake of infants of 1–4 months age of 4.46 g/kg/day [67], it can be assumed that an infant is exposed to approximately 900 ng/kg/day CPs. Given an estimated NOEL or PNEC of 1 mg/kg/day, there appears to be a reasonably large margin of safety of 1,000. This rough estimate of infant exposure, however, comes quite close to the TDI reported by the Government of Canada of 6 μg/kg/day [5].

5   Concluding Remarks

Toxicity studies in rodents and environmental species indicate that chronic exposure to relatively low concentrations of particularly SSCPs and MCCPs adversely affects survival and development. Much of the toxicity data on the CPs is dated, and suitable analytical methods were not available to well characterize internal doses or exposure concentrations. The development of improved analytical methods and monitoring of these heterogeneous compounds is challenging, but has certainly improved over recent years. This is essential in distinguishing the possible effects of impurities in technical CP mixtures and will aid in the accurate comparison of exposure data with hazard information based on internal dose to establish better risk characterization.

More research is clearly needed to understand the mechanisms of toxicity of CPs. While the SSCPs have been implicated as carcinogenic, the mechanisms by which these compounds induce tumours are still unclear. Also, carcinogenicity data for the MCCPs and LCCPs is lacking. Data on the neurotoxicity is limited, while nothing is known of the immunotoxicity of these chemicals. Toxicity data on individual CP congeners is lacking, and is necessary to determine the influence of the chain lengths and degrees of chlorination on toxicodynamics, toxicokinetics and metabolism of CPs. Given the sensitivity of early life stages to CP exposure, in addition to the finding of CPs in breast milk at significant levels, more research is needed to determine the effects of perinatal exposure.

Despite the data gaps, it is clear that, especially for the SSCPs, these compounds may lead to significant adverse environmental and human health effects. Given their persistence in the environment and propensity to bioaccumulate, as well as to undergo long-range environmental transport, it is clear that action regarding the regulation of these compounds is warranted.

References

1. Serrone DM, Birtley RDN, Weigand W, Millischer R (1987) Toxicology of chlorinated paraffins. Food Chem Toxicol 25:553–562
2. UNEP (United National Environment Program) (2008) Updated supporting document for the risk profile on short-chained chlorinated paraffins (UNEP/POPS/POPRC.4/10), Persistent Organic Pollutants Review Committee Fourth meeting, Geneva, 13–17 October 2008, pp 56. http://chm.pops.int/Portals/0/Repository/poprc4/UNEP-POPS-POPRC.4-INF-22.English. DOC. Accessed 3 Nov 2009
3. Rossberg M, Lendle W, Pfleiderer G, Tögel A, Dreher E-L, Langer E (2003) Chlorinated hydrocarbons. Ullmann's encyclopedia of industrial chemistry. Wiley-VCH, Weinheim
4. Nikiforov VA (2010) Synthesis of polychloroalkanes. In: de Boer J (ed) The handbook of environmental chemistry: chlorinated paraffins. Springer, Heidelberg
5. Government of Canada (1993) Priority substances list assessment report: chlorinated paraffins. Ministry of Supply and Services, Canada, p 36
6. WHO (World Health Organization) (1996) Environmental health criteria 181: chlorinated paraffins, Geneva. http://www.inchem.org/documents/ehc/ehc/ehc181.htm. Accessed 3 Nov 2009
7. European Commission (2000) European union risk assessment report. 1st priority list vol 4: alkanes, C_{10-13}, chloro–. European Chemicals Bureau, Luxembourg, pp 166
8. Tomy GT, Fisk AT, Westmore JB, Muir DCG (1998) Environmental chemistry and toxicology of polychlorinated *n*-alkanes. Rev Environ Contam Toxicol 158:53–128
9. Howard PH, Santodonato J, Saxena J (1975) Investigations of selected potential environmental contaminants: chlorinated paraffins. Syracuse University Research Corporation, Syracuse, p 107 (Report No 68-01-3101)
10. NTP (The National Toxicology Program) (1986) NTP technical report on the toxicology and carcinogenesis studies of chlorinated paraffins (C_{12}, 60% chlorine) (CAS No. 63449-39-8) in F344/N Rats and B6C3F1 Mice (Gavage Studies). NTP TR 308, NIH Publication No. 86-2564, Research Triangle Park, NC, USA
11. Bucher JR, Alison RH, Montgomery CA, Huff J, Haseman JK, Farnell D, Thompson R, Prejean JD (1987) Comparative toxicity and carcinogenicity of two chlorinated paraffins in F344/N rats and B6C3F1 mice. Fund Appl Toxicol 9:454–468
12. IRDC (International Research and Development Corporation) (1981) 14-Day oral toxicity study in rats with chlorinated paraffin: 58% chlorination of short chain length n-paraffin. Studies conducted for the Working Party of the Chlorinated Paraffin Manufacturers Toxicology Testing Consortium (Report No 438-006)
13. Ahotupa M, Hietanen E, Mantyla E, Vainio H (1982) Effects of polychlorinated paraffins on hepatic, renal and intestinal biotransformation rates in comparison to the effects of polychlorinated biphenyls and naphthalenes. J Appl Toxicol 2:47–53
14. Elcombe CR, Watson SC, Wyatt I, Foster JR (1994) Chlorinated paraffins (CP): mechanisms of carcinogenesis. Toxicologist 14:276

15. Hallgren S, Darnerud PO (2002) Polybrominated diphenyl ethers (PBDEs), polychlorinated biphenyls (PCBs) and chlorinated paraffins (CPs) in rats: testing interactions and mechanisms for thyroid hormone effects. Toxicology 177:227–243
16. Birtley RDN, Conning DM, Daniel JW, Ferguson DM, Longstaff E, Swan AAB (1980) The toxicological effects of chlorinated paraffins in mammals. Toxicol Appl Pharmacol 54:514–525
17. Wyatt I, Coutts CT, Elcombe CR (1993) The effect of chlorinated paraffins on hepatic enzymes and thyroid hormones. Toxicology 77:81–90
18. Poon R, Lecavalier P, Chan P, Viau C, Håkansson H, Chu I, Valli VE (1995) Subchronic toxicity of a medium-chained chlorinated paraffin in the rat. J Appl Toxicol 15:455–463
19. NTP (The National Toxicology Program) (1986) NTP technical report on the toxicology and carcinogenesis studies of chlorinated paraffins (C23, 43% chlorine) (CAS No 63449- 39-8) in F344IN rats and B6C3F1 mice (Gavage studies). NTP TR 305, NIH Publication No 86-2561, Research Triangle Park, NC, USA
20. IRDC (International Research and Development Corporation) (1981) 14-day oral toxicity study in rats with chlorinated paraffin: 43% chlorination of long chain length n-paraffin. Studies conducted for the Working Party of the Chlorinated Paraffin Manufacturers Toxicology Testing Consortium (Report No 438-005)
21. IRDC (International Research and Development Corporation) (1981) 14-day dietary range-finding study in rats with chlorinated paraffin: 70% chlorination of long chain length nparaffin. Studies conducted for the Working Party of the Chlorinated Paraffin Manufacturers Toxicology Testing Consortium (Report No 438-004)
22. IARC (International Agency for Research on Cancer) (1990) Summaries and evaluations. Chlorinated paraffins (Group 2B), vol 48, p 55
23. Myhr B, McGregor D, Bowers L, Riach C, Brown AG, Edwards I, McBride D, Martin R, Caspary WJ (1990) L5178Y Mouse Lymphoma Cell Mutation Assay Results With 41 Compounds. Environ Mol Mutagen 16(Suppl 18):138–167
24. Anderson BE, Zeiger E, Shelby MD, Resnick MA, Gulati DK, Ivett JL, Loveday KS (1990) Chromosome aberration and sister chromatid exchange test results with 42 chemicals. Environ Mol Mutagen 16(Suppl 18):55–137
25. Kato Y, Kenne K (1996) Inhibition of cell-cell communication by commercial chlorinated paraffins in rat liver epithelial IAR 20 cells. Pharmacol Toxicol 79:23–28
26. Ashby J, Lefevre PA, Elcombe CR (1990) Cell replication and unscheduled DNA synthesis (UDS) activity of low molecular weight chlorinated paraffins in the rat liver in vivo. Mutagenesis 5:515–518
27. IRDC (International Research and Development Corporation) (1983) In vivo cytogenetic evaluation by analysis of rat bone marrow cells. Chlorinated paraffins: 58% chlorination of short chain length n-paraffins. Mattawan, MI, International Research and Development Corporation, pp 58 (Report No 438-013)
28. IRDC (International Research and Development Corporation) (1983) In vivo cytogenetic evaluation by analysis of rat bone marrow cells. Chlorinated paraffin: 52% chlorination of intermediate chain length n-paraffins. International Research and Development Corporation, Mattawan, MI, pp 62 (Report No 438-014)
29. IRDC (International Research and Development Corporation (1983) In vivo cytogenetic evaluation by analysis of rat bone marrow cells. Chlorinated paraffin: 70% chlorination of long chain length n-paraffins. International Research and Development Corporation, Mattawan, MI, pp 70 (Report No 438-016)
30. IRDC (International Research and Development Corporation (1983) In vivo cytogenetic evaluation by analysis of rat bone marrow cells. Chlorinated paraffin: 43% chlorination of long chain length n-paraffins. International Research and Development Corporation, Mattawan, MI, pp 56 (Report No 438-012)
31. Hoechst AG (1989) Chlorowax 500°C micronucleus test in male and female NMRI mice after oral administration. Frankfurt/Main, Germany, p 24 (Report No 89.0253)

32. IRDC (International Research and Development Corporation (1983) Dominant lethal study in rats. Chlorinated paraffin: 58% chlorination of short chain *n*-paraffins. International Research and Development Corporation, Mattawan, MI, pp 54 (Report No 438-011)

33. IRDC (International Research and Development Corporation) (1985) Reproduction range-finding study in rats with chlorinated paraffin: 52% chlorination of intermediate chain length *n*-paraffin. Studies conducted for the Working Party of the Chlorinated Paraffin Manufacturers Toxicology Testing Consortium (Report No 438-049)

34. IRDC (International Research and Development Corporation) (1983) Teratology study in rabbits with chlorinated paraffin: 52% chlorination of intermediate chain length *n*-paraffin. Studies conducted for the Working Party of the Chlorinated Paraffin Manufacturers Toxicology Testing Consortium (Report No 438-032)

35. IRDC (International Research and Development Corporation) (1984) Teratology study in rats with chlorinated paraffin: 52% chlorination of intermediate chain length *n*-paraffin. Studies conducted for the Working Party of the Chlorinated Paraffin Manufacturers Toxicology Testing Consortium (Report No 438-017)

36. IRDC (International Research and Development Corporation) (1982) Teratology study in rabbits with chlorinated paraffin: 70% chlorination of long chain length *n*-paraffin. Studies conducted for the Working Party of the Chlorinated Paraffin Manufacturers Toxicology Testing Consortium (Report No 438-039)

37. IRDC (International Research and Development Corporation) (1983) Teratology study in rabbits with chlorinated paraffin: 58% chlorination of short chain length *n*-paraffin. Studies conducted for the Working Party of the Chlorinated Paraffin Manufacturers Toxicology Testing Consortium (Report No 438-031)

38. IRDC (International Research and Development Corporation) (1983) Teratology study in rats with chlorinated paraffin: 43% chlorination of long chain length *n*-paraffin. Studies conducted for the Working Party of the Chlorinated Paraffin Manufacturers Toxicology Testing Consortium (Report No 438-015)

39. IRDC (International Research and Development Corporation) (1981) Teratology study in rabbits with chlorinated paraffin: 43% chlorination of long chain length *n*-paraffin. Studies conducted for the Working Party of the Chlorinated Paraffin Manufacturers Toxicology Testing Consortium (Report No 438-030)

40. IRDC (International Research and Development Corporation) (1983) Teratology study in rats with chlorinated paraffin: 70% chlorination of long chain length *n*-paraffin. Studies conducted for the Working Party of the Chlorinated Paraffin Manufacturers Toxicology Testing Consortium (Report No 438-045)

41. Darnerud PO, Lundkvist U (1987) Studies on implantation and embryonic development in mice given a highly chlorinated hexadecane. Pharmacol Toxicol 60:239–240

42. Eriksson P, Kihlström J (1985) Disturbance of motor performance and thermoregulation in mice given two commercial chlorinated paraffins. Bull Environ Contam Toxicol 34:205–209

43. Eriksson P, Nordberg A (1986) The effects of DDT, DDOH-palmitic acid, and a chlorinated paraffin on muscarinic receptors and the sodium-dependent choline uptake in the central nervous system of immature mice. Toxicol Appl Pharmacol 85:121–127

44. Thompson RS, Madeley JR (1983) Toxicity of a chlorinated paraffin to the marine algae *Skeletonema costatum*. Imperial Chemical Industries PLC, Devon, England (Brixham Report No BL/B/2328)

45. Thompson RS, Madeley JR (1983) Toxicity of a chlorinated paraffin to the green algae *Selenastrum capricornutum*. Imperial Chemical Industries PLC, Devon, England (Brixham Report No BL/B/2321)

46. Thompson RS, Madeley JR (1983) The acute and chronic toxicity of a chlorinated paraffin to *Daphnia magna*. Imperial Chemical Industries PLC, Devon, England (Brixham Report No BL/B/2358)

47. Thompson RS, Madeley JR (1983) The acute and chronic toxicity of a chlorinated paraffin to the mysid shrimp *(Mysidopsis bahia)*. Imperial Chemical Industries PLC, Devon, England (Brixham Report No BL/B/2373)

48. Thompson RS, Shillabeer N (1983) Effect of a chlorinated paraffin (58% chlorination of short chain length *n*-paraffins) on the growth of mussels (*Mytilus edulis*). Imperial Chemical Industries Ltd, Brixham Laboratory, Brixham, p 54 (Report No BL/B/2331)
49. Madeley JR, Thompson RS (1983) Toxicity of chlorinated paraffin to mussels (*Mytilus edulis*) over 60 days. Imperial Chemical Industries PLC, Devon, England (Brixham Report No BL/B/2289)
50. Madeley JR, Maddock BG (1983) Toxicity of a chlorinated paraffin to rainbow trout over 60 days. Imperial Chemical Industries PLC, Devon, England (Brixham Report No BL/B/2202)
51. EG & G Bionomics (1983) The acute and chronic toxicity of a chlorinated paraffin (58% chlorination of short chain length *n*-paraffins) to midges (*Chironomus tentans*). EG & G Bionomics, Wareham, MA, pp 39 (Report No BW 83-6-1426)
52. Bengtsson BE, Baumann Ofstad E (1982) Long-term studies on uptake and elimination of some chlorinated paraffins in the bleak, Alburnus alburnus. Ambio 11:38–40
53. Bengtsson BE, Svanberg O, Lindén E, Lunde G, Baumann Ofstad E (1979) Structure related uptake of chlorinated paraffins in bleaks (*Alburnus alburnus L*). Ambio 8:121–122
54. Hill RW, Maddock BG (1983) Effect of a chlorinated paraffin (58% chlorination of short chain length *n*-paraffins) on embryos and larvae of the sheepshead minnow (*Cyprinodon variegatus*)-(study one). Imperial Chemical Industries Ltd, Brixham Laboratory, Brixham, p 57 (Report No BL/B/2326)
55. Haux C, Larsson Å, Lidman U, Förlin L, Hansson T, Johansson-Sjöbeck M-L (1982) Sublethal physiological effects of chlorinated paraffins on the flounder, *Platichtys flesus L*. Ecotoxicol Environ Saf 6:49–59
56. Cooley HM, Fisk AT, Wiens SC, Tomy GT, Evans RE, Muir DC (2001) Examination of the behavior and liver and thyroid histology of juvenile rainbow trout (Oncorhynchus mykiss) exposed to high dietary concentrations of C(10)-, C(11)-, C(12)- and C(14)-polychlorinated *n*-alkanes. Aquat Toxicol 54:81–99
57. Lindén E, Bengtsson BE, Svanberg O, Sundström G (1979) The acute toxicity of 78 chemicals and pesticide formulations against two brackish water organisms, the bleak *(Alburnus alburnus)* and the harpacticoid *Nitocra spinipes*. Chemosphere 8:843–851
58. Madeley J, Birtley R (1980) Chlorinated paraffins and the environment. 2. Aquatic and avian toxicology. Environ Sci Technol 14:1215–1221
59. Madeley JR, Maddock BG (1983) Effects of a chlorinated paraffin on the growth of rainbow trout. Chlorinated paraffin: 58% chlorination of short chain length paraffins. Imperial Chemical Industries Ltd, Brixham Laboratory, Brixham, UK, p 67 (Report No BL/B/2309)
60. Bezchlebová J, Cernohlávková J, Kobeticová K, Lána J, Sochová I, Hofman J (2007) Effects of short-chain chlorinated paraffins on soil organisms. Ecotoxicol Environ Saf 67:206–211
61. Sochová I, Hofman J, Holoubek I (2007) Effects of seven organic pollutants on soil nematode *Caenorhabditis elegans*. Environ Intern 33:798–804
62. Brunström B (1983) Toxicity in chick embryos of three commercial mixtures of chlorinated paraffins and of toxaphene injected into eggs. Arch Toxicol 54:353–357
63. Brunström B (1985) Effects of chlorinated paraffins on liver weight, cytochrome P-450 concentration and microsomal enzyme activities in chick embryos. Arch Toxicol 57:69–71
64. Ueberschar KH, Matthes S (2004) Dose–response feeding study of chlorinated paraffins in broiler chickens: effects on growth rate and tissue distribution. Food Addit Contam 21:943–948
65. Ueberschär KH, Dänicke S, Matthes S (2007) Dose-response feeding study of short chain chlorinated paraffins (SCCPs) in laying hens: effects on laying performance and tissue distribution, accumulation and elimination kinetics. Mol Nutr Food Res 51(2):248–254
66. Thomas GO, Farrar D, Braekevelt E, Stern G, Kalantzi OI, Martin FL, Jones KC (2006) Short and medium chain length chlorinated paraffins in UK human milk fat. Environ Intern 141:30–41
67. Butte NF, Wong WW, Ferlic L, Smith EO, Klein PD, Garza C (1991) Energy expenditure and deposition of breast-fed and formula-fed infants during early infancy. Pediatr Res 28:631–640

Risk Assessment of Short-Chain Chlorinated Paraffins in Japan

Kiyotaka Tsunemi

Abstract The objective of this chapter is to assess the ecological and human health risk of short-chain chlorinated paraffins (SCCPs) via the environment in Japan. First, release sources of SCCPs are identified based on the substance flow analysis, and the volume of their releases is estimated. Next, the behavior of SCCPs in the environment is estimated using a multimedia model, and the estimated SCCPs concentrations in the environment and food are validated with measured concentrations. Then, the endpoints and doses of SCCPs as the criteria for ecological and human health risk assessment are derived through the review of the existing toxicological data. Finally, risk characterization is performed based on the results of exposure assessment and dose-response assessment. As a result, it is determined that there is little need to be concerned about potential ecological risk to aquatic, sediment-dwelling, and soil-dwelling organisms in local areas and the regions in Japan. As far as human health risk is concerned, the Margins of exposure are 1.5×10^5 and 2.2×10^6, which are larger than uncertainty factors. Thus, it is determined that there is no significant human health risk via the environment.

Keywords Ecological risk, Exposure assessment, Human health risk, Multimedia model, Risk assessment, Short-chain chlorinated paraffins

Contents

K. Tsunemi
National Institute of Advanced Industrial Science and Technology, Research Institute of Science for Safety and Sustainability, Onogawa 16-1, Tsukuba, Ibaraki, Japan
e-mail: k-tsunemi@aist.go.jp

J. de Boer (ed.), *Chlorinated Paraffins*, Hdb Env Chem (2010) 10: 155–194,
DOI 10.1007/698_2009_35, © Springer-Verlag Berlin Heidelberg 2010,
Published online: 12 January 2010

Abbreviations

BCF	Bioconcentration factor
CPs	Chlorinated paraffins
EA	Environment agency
EUSES	European union system for the evaluation of substances
FRCJ	Flame retardant chemicals association of Japan
HC_5	Hazard concentration 5
LCCPs	Long-chain chlorinated paraffins
LOD	Limit of detection
MCCPs	Medium-chain chlorinated paraffins
MHLW	Ministry of health, labor and welfare
MLIT	Ministry of land, infrastructure and transport
MOE	Margin of exposure
NOAEL	No observed adverse effect level
NOEC	No observed effect concentration
NOEL	No observed effect level
NTP	National toxicology program
PNEC	Predicted no effect concentration
PRTR	Pollutant release and transfer register
SCCPs	Short-chain chlorinated paraffins
SRTI	Statistical research and training institute
SS	Suspended solid
SSD	Species sensitivity distribution
TGD	Technical guidance document
TRI	Toxics release inventory

1   Introduction

In Japan, no regulation for short-chain chlorinated paraffins (SCCPs) had been implemented and SCCPs are not included in the list of designated chemical substances for the Pollutant Release and Transfer Register (PRTR). Risk assessment of SCCPs, which was performed in Europe, the U.S, Canada, and Australia, was not conducted in Japan, and the releases of SCCPs into the environment, which can be obtained from the Toxics Release Inventory (TRI) in the U.S., are not available.

In 2003, the Law Concerning the Evaluation of Chemical Substances and Regulation of Their Manufacture, etc. was amended in Japan. The main aims of the amendment are the implementation of the system to assess and regulate chemical substances from the viewpoint of the effects on plants and animals in the environment, the introduction of regulations for existing chemical substances with persistence and high bioconcentration, and a screening system based on the probabilities of releases into the environment. In February 2005, SCCPs were classified into Class I Chemical Substances Monitored, which include substances with persistence and high bioconcentration. Therefore, public attention on the risk of SCCPs has been increased.

With this background, National Institute of Advanced Industrial Science and Technology (AIST) conducted a risk assessment for SCCPs with regard to the domestic industrial structures related to SCCPs including production, use, and waste management processes [1].

2   Releases into the Environment

With the current regulatory situation in Japan, where no regulation of SCCPs has been implemented and SCCPs are not designated as PRTR-list substances, almost no information is available on their production and use to identify release resources, and also very little data of environmental monitoring have been obtained. Under such difficult conditions with limited data on SCCPs, the demand and use of SCCPs are estimated in this assessment not only using statistics and reported data but also by obtaining information via interviews with companies and industrial organizations; further releases into the environment are estimated using release factors established overseas as required.

2.1   Substance Flow Analysis

SCCPs are used as extreme-pressure additive to metal working fluids, flame retardant, and plasticiser of plastics, rubber, and paints, and the products including these additives are used for various purposes. As shown in Fig. 1, the usage and

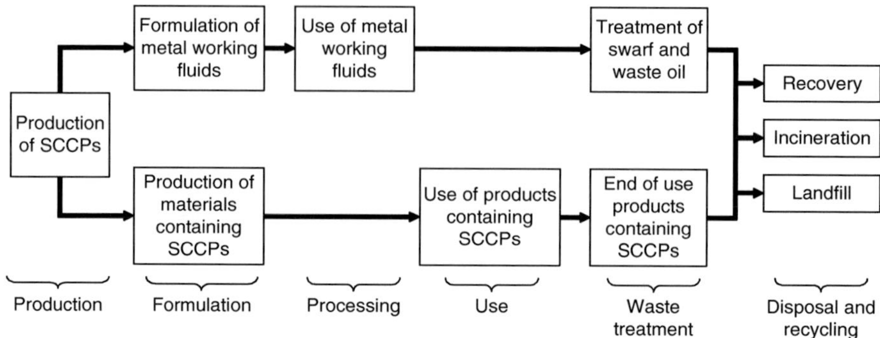

Fig. 1 Main life cycle stages of SCCPs

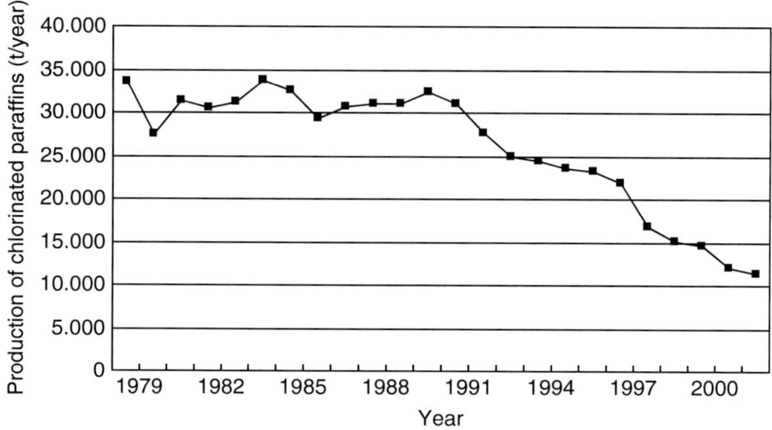

Fig. 2 Interannual changes in production of total CPs [2]

releases into the environment of SCCPs are estimated on each life cycle stage of SCCP production, formulation, processing, use, waste treatment, disposal, and recycling. This study estimated the substance flow at each of these stages in Japan in 2001.

Interannual changes in production of total chlorinated paraffins (CPs) reported by the Flame Retardant Chemicals Association of Japan (FRCJ) [2] are shown in Fig. 2. FRCJ also reported the production and shipment of SCCPs, medium-chain chlorinated paraffins (MCCPs), and long-chain chlorinated paraffins (LCCPs) in 2001 [2] as shown in Table 1. It indicates that ≈500 tons per year of SCCPs were shipped, suggesting that SCCPs were not used for polyvinyl chloride. In the Kanto region, 500 tons per year of SCCPs were mainly used in oil fluids. In contrast, nearly 200 tons of SCCPs were shipped to the Kansai region for other uses including flame retardant additives and lipid additives for rubber, paint, and adhesives. As shown in Table 1, approximately half of SCCPs were used in oil fluids and the other half for other products.

Table 1 Domestic shipment of CPs in 2001 by region [2]

Product	Region	Use for oil fluid (tons per year)	Use for polyvinyl chloride (tons per year)	Other use (tons per year)	Total (tons per year)
SCCPs	Kanto	175	–	32	207
	Kansai	31	–	190	221
	Overall Japan	240	–	262	502
MCCPs	Kanto	2,279	1,516	348	4,143
	Kansai	710	1,951	173	2,834
	Overall Japan	3,199	3,619	799	7,617
LCCPs	Kanto	190	592	1,347	2,129
	Kansai	145	243	618	1,006
	Overall Japan	390	966	2,365	3,721
Total CPs	Kanto	2,644	2,108	1,727	6,479
	Kansai	886	2,194	981	4,061
	Overall Japan	3,829	4,585	3,426	11,840

The amount of MCCPs was the largest among total CPs and 7,600 tons of MCCPs were shipped all over Japan, mainly for use as plasticizers for polyvinyl chloride. Also, 3,700 tons of LCCPs were shipped mainly for other uses. Based on these data, the production percentage of SCCPs to total CPs is estimated to be 4.15% in 2001.

The above data are the only available data on SCCP production and no specific data before 2000 are obtained. Therefore, interannual changes of SCCP production are estimated, based on the assumption that the percentage of 4.15% for SCCPs was constant from the beginning of production in 1950. Specifically, it is assumed that SCCP production was increased in proportion to the enhanced production of total CPs along with the industrial development in Japan, and the promotion of substitution with non-chlorine-based products had similar effects, both on SCCPs and total CPs. Also, it is assumed that SCCP production was increased linearly during the period from 1950 to 1975 and 1975 to 1979 because of the lack of data. Further, it is assumed that 50% of SCCPs were used in metal working fluids and another 50% for other products including flame retardant additives and lipid additives. As a result, the estimated consumptions of SCCPs from 1950 to 2002 in Japan are obtained and presented in Fig. 3, which indicate a profile with a peak of 1,200–1,400 tons per year from 1980 to 1990, and a decrease to 500 tons per year by 2000.

The substance flow of SCCPs in 2001 is estimated as shown in Table 2. As of 2002, there are three domestic companies (four plants) producing CPs; Tosoh (Shin-nanyo and Sakata, total production capacity: 18,000 tons per year), Ajino-moto Fine-Techno (Kawasaki, production capacity: 12,000 tons per year), and ADEKA (Kajima, production capacity: 12,000 tons per year) [3]. Two plants with similar production capacity are located in the Kanto region and their production capacity was ≈50–60% of all plants in Japan as of 2000. When the domestic distribution of production is almost equal to the production capacity, it is assumed that 50% of nationwide production is in the Kanto region, of which 50% is produced in one plant, and used as the basis for exposure analysis.

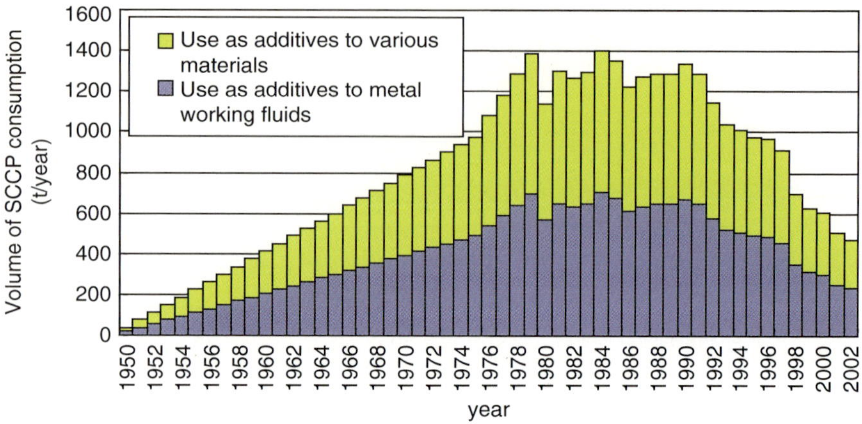

Fig. 3 Estimated consumption of SCCPs from 1950 to 2002 in Japan

Table 2 Volumes of production and use of SCCPs in 2001. Figures in parenthesis indicate the range from half to twice of the median

Life cycle stage	Whole Japan (tons per year)	Kanto region (tons per year)	Plant as a local area (tons per year)
SCCP production	502	250	125
Metal working fluid formulation	240	175	35
Metal working fluid use	240	72	7.2 (3.6–14.4)
Manufacturing of SCCP- containing products	262	32	3.2 (1.6–6.4)

Based on the reliable data provided by FRCJ as shown in Fig. 4 [2], the SCCPs in metal working fluids in the stages of production and the use of metal working fluids is determined to be 240, of which 175 tons were used in the Kanto region. In the Kanto region, more than five formulation plants for metal working fluids are located, and therefore, it is assumed that 20% of the usage of the Kanto region is used in each plant.

The usage of metal working fluids containing SCCPs in the Kanto region is estimated from the values of shipment related to metal working in nationwide industrial statistics. As the shipment of Kanto region was ≈30% of the total values in Japan from the industrial statistics in 2001, it is assumed that the volumes of fluid usage are in proportion to the shipment values of 240 tons of SCCPs used in fluids across the country, ≈30%; 72 tons is estimated to have been used in the Kanto region. There are, however, no data available on the number of companies using metal working fluids containing SCCPs. Although many small companies are located, assuming that ten large and middle companies are located in the Kanto region, the scenario is set for 10% of usage of the Kanto region being used in each plant.

The total volume of SCCPs in plastics, gum, paint, and adhesive is determined to be 262 tons without classifying the volume of each use of SCCPs.

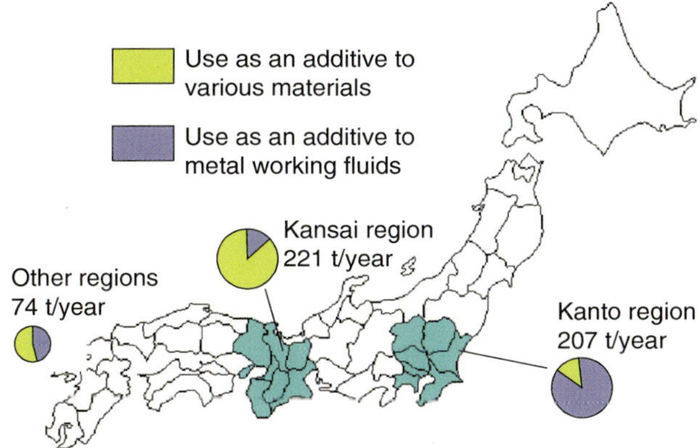

Fig. 4 Amount of SCCPs demanded by regions in 2001 [2]

The SCCP use for manufacturing SCCP-containing products in the Kanto region is estimated to be 32 tons [2], and $\approx$10% of usage of the Kanto region is used in each plant, similar to the use of metal working fluids.

To estimate the environmental concentration in a local area near the release site, the usage of SCCPs in one plant in the Kanto region is estimated. Firstly, it is required to estimate the number of plants that are located in the Kanto region. SCCPs are used only for specific purposes including additives to metal working fluids for difficult metal processes and flame-retardant additives, and thus, plants using SCCPs are expected to be limited. It is assumed that one plant using SCCPs is located around a sewage treatment plant with the average capacity in Japan and releases SCCPs into water through the sewage treatment plant. In the stages of SCCP production and metal working fluid formulation, the approximate number of plants in the Kanto region is estimated. However, in the stages of using metal working fluids and manufacturing SCCP-containing products, the number of plants is estimated based on insufficient data. Therefore, a sensitivity analysis is to be conducted using the estimated usage as a median and with the range from half to twice the median.

The amount of end-use products containing SCCPs was estimated using the following equation that represents Weibull distribution that is a continuous probability distribution and is often used in the field of life data analysis due to its flexibility.

$$f(x) = \left(\frac{u}{v^u}\right)x^{u-1}\exp\left[-\left(\frac{x}{v}\right)^u\right] \tag{1}$$

where $f(x)$ is the function of lifetime, u and v are the parameters of the Weibull distribution. Because the ratio of SCCP-containing products is unclear, the uses of CPs are combined. Assuming that 90% of final products are disposed of in 5–30 years, parameters are estimated as $u = 3$ and $v = 17.5$. The interannual changes in SCCP production that are shown in Fig. 3 are applied to this function; the cumulative volumes of SCCP production are estimated. As a result, the cumulative disposal from 1950 to 2001 is estimated to be $\approx$15,800 tons and the volume of stock of SCCPs in 2001, as contained in products, is estimated to be 5,300 tons. Assuming that the disposal of SCCP-containing products depends on population distribution, the cumulative disposal of SCCPs in the Kanto region is estimated to be 30% of the total volume in Japan, i.e., 4,740 tons.

2.2 Estimation of Releases into the Environment

Although there are no release data for SCCPs in Japan, relevant guidance documents and risk assessment reports were referred to derive appropriate release factors [4] as shown in Table 3.

Multiplying the estimated volumes of SCCP use by each of the SCCP release factors by life cycle stage, the SCCP releases into the environment are estimated as shown in Table 4. Regarding the volumes of local releases, a 10% release factor from a sewage treatment plant into river water is applied to the releases from a single plant in the Kanto region to estimate the release after waste treatment.

Of the total volume of SCCPs released into water, it is assumed that 63.5%, which is the ratio of sewage system coverage in 2001, flows into sewage treatment plants. Total SCCP releases from life cycle processes into water (whole Japan: 19,095 kg per year, Kanto region: 5,944 kg per year) are multiplied by the treatment percentage in domestic sewage treatment plants (63.5%) [5], and as a result, the volumes of SCCPs in influent into sewage treatment plants are estimated to be 12,125 kg per year (whole Japan) and 3,774 kg per year (Kanto region). Subsequently, the direct releases of SCCPs in surface water are estimated to be 6,970 kg per year (whole Japan) and 2,170 kg per year (Kanto region). These values are multiplied by the release factor into water (10%) and transfer factor into sludge

Table 3 Release factors of SCCPs by life cycle stage

Life cycle stage	Air release factor (%)	Water release factor (%)	Soil release factor (%)	Source
Production of SCCPs	0	0.01	0	[37]
Formulation of metal working fluids	0.005	0.2	0.001	[40]
Use of metal working fluids	8	5	0	[1]
Waste treatment of scraps after use of metal working fluids	0	0.82	0	[1]
Disposal of waste oil after use of metal working fluids	0	0.039	0	[1]
Manufacturing of SCCP-containing products	0	0.001	0	[37]
Use of SCCP-containing products (outdoor)	0.0029	0.16	0	[42]
Use of SCCP-containing products (indoor)	0.0029	0.0029	0	[42]
Waste treatment of SCCPs	0	0.0029	0	[42]

Table 4 Estimated SCCP releases in 2001. Figures in parenthesis indicate the range from a half to twice of the median

Life cycle stage	Releases in the whole of Japan (kg per year)	Releases in the Kanto region (kg per year)	Releases in a local area (kg per year)	Medium to which release occurs
Production of SCCPs	50	25	13	Water
Formulation of metal working fluids	12	8.75	1.75	Air
	480	350	70	Water
	2.4	1.75	0.35	Soil
Use of metal working fluids	19,200	5,760	576 (288–1,152)	Air
	12,000	3,600	360 (180–720)	Water
Waste treatment of scraps after use of metal working fluids	1,968	590		Water
Disposal of waste oil after use of metal working fluids	93.6	28.1		Water
Manufacturing of SCCP-containing products	2.6	0.32	0.03 (0.02– 0.06)	Water
Use of SCCP-containing products	154	46		Air
	4,317	1,295		Water
Waste treatment of SCCP-containing products	183	55		Water
Total	19,366	5,815		Air
	19,095	5,944		Water
	2.4	1.75		Soil

Table 5 Estimated SCCP releases from sewage treatment plants in 2001

	Japan	Kanto region
Release into surface water (kg per year)	1,213	377
Release into sludge (kg per year)	10,913	3,397

(90%) at sewage treatment plants and the SCCP releases from sewage treatment plants are estimated as shown in Table 5.

The flow of SCCPs through sewage treatment plants is summarized as shown in Fig. 5. It is estimated that the volumes of SCCPs released in agricultural land are 218 kg per year in the whole of Japan and 3.40 kg per year in the Kanto region, respectively from the values in Table 5.

In this section, the release sources of SCCPs into the various environmental compartments in 2001 are identified and shown in Fig. 6. The material flow of SCCPs consists of two major routes of metal working fluids and other products containing SCCPs, and each route has life cycle stages of production, use, and disposal. SCCPs are released mainly at the use stage of metal working fluids into the air (19,200 kg per year) and water (12,000 kg per year), which is expected to have local effects in the areas around metal working plants. The second largest is the release into water from the use stage of SCCP-containing products (4,317 kg per year), which is estimated from the accumulated volume of SCCPs in the products

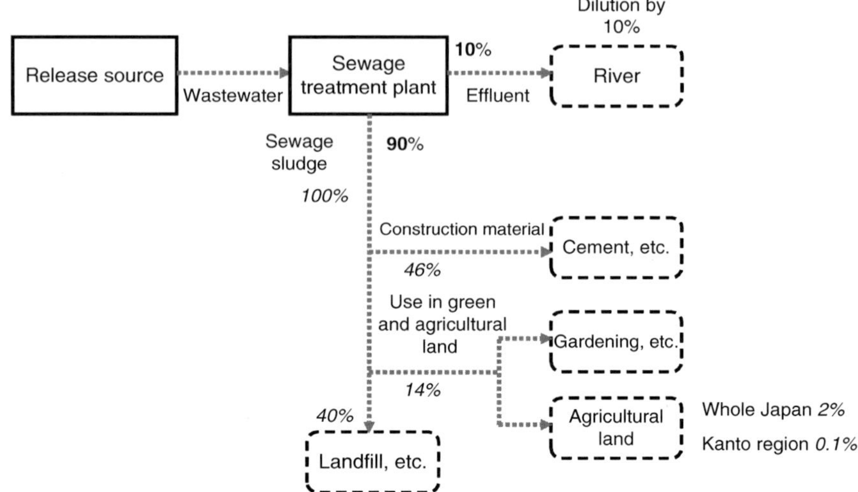

Fig. 5 Estimated flow of SCCPs through sewage treatment plant. The figures in italics indicate the percentages of the volume of SCCPs by each recycling stage to that of SCCPs in sewage sludge

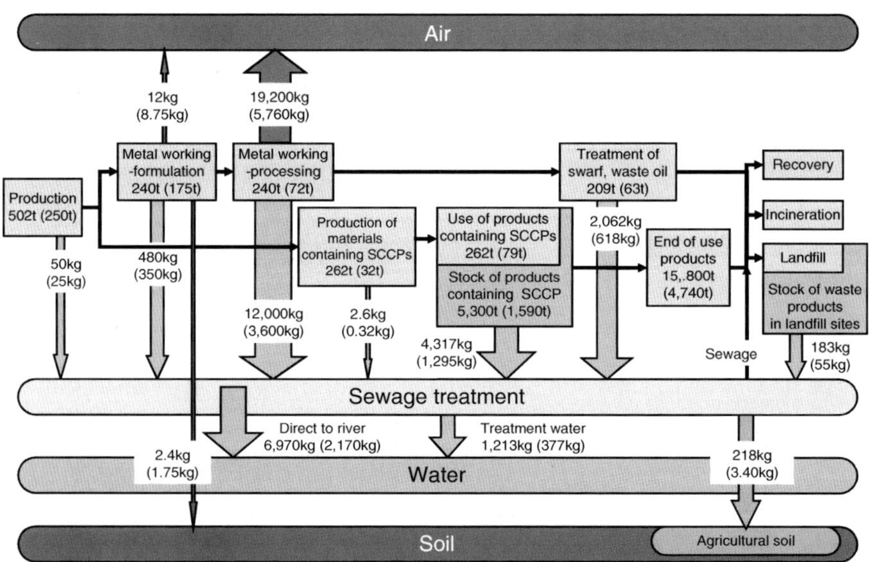

Fig. 6 Flow and emission of SCCPs in 2001 in Japan and Kanto region. The figures in parenthesis indicate the flow and emission in Kanto region

shipped up to 2001. As these products were dispersed and had little local effect, it is relevant to assess their effects only with spatial levels of the Kanto region and the whole of Japan.

3 Exposure Assessment

SCCPs released into the environment reach humans and organisms through various pathways. It is necessary to identify the main exposure routes from release sources to humans and organisms in the environment in order to assess exposures to SCCPs. In this section, the existing data about equilibrium partition and degradation of SCCPs are summarized. Also, the behavior of SCCPs in environmental media including air and soil and the biokinetics in plants and fish is estimated using a multimedia model, and the main exposure routes are identified. The concentrations of SCCPs in the environment and food that were obtained in domestic monitoring are reviewed, and the observed data, mainly in the Kanto region where high volumes of SCCPs are used, are summarized. Based on these results, the concentrations of SCCPs in the environment and food are identified for risk assessment.

3.1 Behavior in Environment

In this section, selecting $C_{12}H_{20}Cl_6$ (chlorine: 56.5%) with C_{12} (mean of C_{10-13}) and Cl_6 (mean of Cl_{1-12}) as the representative SCCP, the parameters used for estimating the behavior of SCCPs in the environment are described briefly.

3.1.1 Physical Properties and Partition Coefficients

The data of physical property used for model application are as follows: molecular weight of $C_{12}H_{20}Cl_6$ is 377 g mol^{-1}, melting point is $-30°C$ [6], vapor pressure is 0.000227 Pa (25°C) [7], and solubility in water is 0.0625 mg L^{-1} (25°C) [8]. The data of partition coefficients used for model application are as follows: log octanol-water partition coefficient (log K_{OW}) is 6.4 [9] and log organic carbon-water partition coefficient (log K_{OC}) is 5.3 [10].

3.1.2 Degradation

Abiotic degradation of SCCPs in the air has been reported [11, 12] estimated that the second reaction rate constant of SCCPs with C_{10-13} and 49–71% chlorine with OH radical was 2.2–8.2 $\times$ 10^{-12} cm^3 $molecule^{-1}$ s^{-1}. The author reported that the half-life in air was estimated to be 1.9–7.2 days, based on the assumption that the OH radical concentration in the air was 5 $\times$ 10^5 molecule cm^{-3}. In this assessment, the half-life as a parameter for modeling is set to be 3.1 days, which is estimated by multiplying the median of the second reaction rate of 5.2 $\times$ 10^{-12} cm^3 $molecule^{-1}$ s^{-1} by the OH radical concentration of 5 $\times$ 10^5 molecule cm^{-3}.

166   K. Tsunemi

Abiotic degradation in other media has not been reported and most available publications are on biodegradation in aerobic conditions. In all of the biodegradation studies, however, the concentrations of SCCPs exceeded the maximum water solubility of SCCPs (0.975 mg L^{-1}). Therefore, it is assumed as a worst case scenario that SCCPs are not degraded in water, soil, sediments, or sewage sludge.

3.1.3 Bioconcentration

Because log K_{OW} of SCCPs is estimated to be 6.4, it is necessary to discuss the bioconcentration of SCCPs, not just for the effects on the survival and reproduction of aquatic organisms but also for those on birds and mammals, the predictors of fish. Furthermore, SCCPs were classified as chemicals with high bioconcentration under the Law Concerning the Examination of Chemical Substances and Regulation of Manufacture, etc.

At the meeting of the Chemical Substances Council held by the Japanese government in 2004, it was reported that a bioconcentration study was conducted in 2003 with SCCPs with C_{11} and $Cl_{7, 8, 9, 10}$ (chlorine: 62.5, 65.7, 68.5 and 70.9%) containing 1% of stabilizer and the quantitative data on each individual substance were obtained [13]. The study was conducted with flow-through system at concentrations of 1 and 0.1 μg L^{-1} of SCCPs, and 20 mg L^{-1} of 2-metoxyethanol was used as dispersant. Liquid chromatography/mass spectrometry (LC/MS) was used for the analysis of the test species, carp (*Cyprinus carpio*). After 62-day exposure, a 14-day excretion study was also conducted. The result is shown in Table 6.

The results indicated that the Bioconcentration factor (BCF) of the whole body at a concentration of 0.1 μg L^{-1} peaked at 38 days of exposure and the BCFs of CPs with C_{11} and $Cl_{7, 8, 9, 10}$ were 5,900, 7,000, 9,500, and 6,700, respectively. BCFs at a

Table 6 Summary of the study on bioconcentration of carp, Cyprinus carpio [13]

CPs	Concentration	BCF, whole body	BCF, skin	BCF, head	BCF, viscera	BCF, edible part
$C_{11}H_{17}Cl_7$	0.1 μg L^{-1}	1,900–5,900	<720	3,900	<720	1,900
				7,000		2,300
$C_{11}H_{17}Cl_7$	1.0 μg L^{-1}	5,400	2,400	5,300	7,400	1,500
			7,800	11,000	20,000	2,900
$C_{11}H_{16}Cl_8$	0.1 μg L^{-1}	2,500–7,000	4,800	6,200	<640	1,600
			5,300	11,000	28,000	4,100
$C_{11}H_{16}Cl_8$	1.0 μg L^{-1}	6,700	2,900	8,700	11,000,	1,800
			14,000	15,000	26,000	3,800
$C_{11}H_{15}Cl_9$	0.1 μg L^{-1}	2,700–9,500	<680	5,500	<680	<680
			<680	12,000	21,000	4,600
$C_{11}H_{15}Cl_9$	1.0 μg L^{-1}	8,100	3,600,	9,500	15,000	2,200
			17,000	18,000	34,000	4,500
$C_{11}H_{14}Cl_{10}$	0.1 μg L^{-1}	2,600–6,700	<620	3,100	<620	<620
			<620	11,000	13,000	<620

Exposure 62 days and depuration 14 days, flow-through test system, dispersant: 2-metoxyethanol, 20 mg L^{-1}

concentration of 1 $\mu g\ L^{-1}$ of BCFs of CPs with C_{11} and $Cl_{7,\ 8,\ 9,\ 10}$ were 8,100, 9,200, 11,000, and 9,200, respectively. With the local analyses of two carps at a concentration of 0.1 $\mu g\ L^{-1}$, BCF ranged from 620 to 720 or less in the skin, 3,100–12,000 in the head, 3,100–28,000 in the viscera, and 1,100–4,700 in the edible part, respectively. The excretion half-life was longer with CPs of higher chlorine, i.e., half-lives of CPs with C_{11} and $Cl_{7,\ 8,\ 9,\ 10}$ at a concentration of 0.1 $\mu g\ L^{-1}$ were 2.6, 6.2, 8.6, and 14 days, respectively, and those at a concentration of 1 $\mu g\ L^{-1}$, 6.7, 10, 9.7, and 13 days, respectively.

The EU risk assessment adopted 7,816, the maximum BCF, which was estimated in a study of SCCPs with C_{10-12} and 58% chlorine in rainbow trout by [14, 15]. In this assessment, considering the high applicability of the data obtained in the bioaccumulation study through the gill of SCCPs conducted in Japan for domestic risk assessment, the BCF through gill in fish is established to be 5,900, BCF of CP with Cl_7, which is most similar to the representative SCCP, $C_{12}H_{20}Cl_6$.

3.2 Estimation of the Main Exposure Routes by Environmental Fate Modeling

The concentrations of SCCPs in the environment are estimated in the local areas around release sources, in Kanto region, and the whole of Japan using the EUSES [16]. The quantities used as inputs in the model are the total amount released in the Kanto region and across Japan. Details of the parameters for land use are shown in Table 7. The average temperature and the amount of rainfall are identified as 15°C and 1,500 mm per year, respectively [17].

The concentrations of SCCPs in the environment estimated by the multimedia model are shown in Tables 7 and 8. The environmental concentrations in Kanto region are higher than the average concentrations in Japan. Further, as shown in Table 3, SCCP concentrations in water and sediments around release sources are higher than those in Kanto region, suggesting that SCCPs remain locally and affect

Table 7 Parameters for land use as used in the model

Scale	Parameter		Value	Year of data
Japan	Area		372,837 km^2	2000
	Number of inhabitants		126,926,000	2000
	Area fraction	Agricultural soil	13.00%	2000
		Natural soil	67.10%	2000
		Water	3.60%	2000
		Industrial and urban soil	16.30%	2000
Kanto region	Area		32,423 km^2	2000
	Number of inhabitants		40,434,000	2000
	Area fraction	Agricultural soil	20.40%	1998–2002
		Natural soil	44.10%	1998–2002
		Water	5.00%	1998–2002
		Industrial and urban soil	30.30%	1998–2002

Table 8 Estimated SCCP concentrations in the environment in Kanto region and Japan

Item	Unit	Kanto region	Japan
Concentration in air	$ng\ m^{-3}$	0.430	0.180
Concentration in water	$\mu g\ L^{-1}$	0.0375	0.0125
Concentration in sediment	$mg\ kg^{-1}$ ww	0.286	0.0959
Concentration in soil	$mg\ kg^{-1}$ ww	0.150	0.0627
Concentration in agricultural soil	$mg\ kg^{-1}$ ww	0.151	0.0790

Table 9 Estimated SCCP concentrations in the environment around release sources

Item	Unit	Production of SCCPs	Production of metal working fluids	Use of metal working fluids	Manufacturing of SCCP-containing products
Concentration in water	$\mu g\ L^{-1}$	0.0567	0.126	0.492 (0.265–0.947)	0.0375 (0.0375–0.0375)
Concentration in sediment	$mg\ kg^{-1}$ ww	0.246	0.629	2.56 (1.36–4.96)	0.163 (0.163–0.163)

Values in parenthesis are the range of concentrations when SCCP use volumes are set in the range of half to twice of the median

Table 10 Estimated SCCP concentrations in food

Food	Concentration
Leaves of plant	2.06 $\mu g\ kg^{-1}$
Root tissue of plant	738 $\mu g\ kg^{-1}$
Meat	13.4 $\mu g\ kg^{-1}$
Milk	4.23 $\mu g\ kg^{-1}$
Fish	221 $\mu g\ kg^{-1}$ ww

the ecosystem. Of these concentrations, the highest concentrations are those in water and sediment from the use stage of metalworking fluids, and the main route is the release into water from metal working plants through sewage facilities (Table 9).

The concentrations of SCCPs in the environment estimated by the multimedia model are shown in Table 10. Concentrations in the root tissue of plants are relatively high. The SCCP concentrations in fish are also high, which, on the other hand, is due to high bioaccumulation of SCCPs.

3.3 Monitoring Data in Japan

While monitoring data of total CPs and LCCPs are available, no data of SCCPs alone are available in Japan. In order to supplement the information on concentrations of SCCPs in the domestic environment by obtaining the measured concentrations of SCCPs in Japan, AIST conducted its own monitoring in the Kanto and Kansai regions, which have many industrial plants and large populations.

3.3.1 Water

In 1980 and 1981, the first monitoring of CPs in the domestic environment was performed by Environment Agency (EA). The environment monitoring for water and sediment was conducted in 1980 [18] and the comprehensive environmental monitoring for water, sediment, and fish (edible part) in 1981 [19]. In these monitorings, the total CPs including SCCPs, MCCPs, and LCCPs were analyzed.

Water samples were collected from 17 sites (three samples/site) in the environment monitoring and 40 sites (three samples/site) in the comprehensive environmental monitoring. CP concentrations were analyzed using GC/ECD and the detection limit in water was 0.01 μg mL^{-1}. In the environment monitoring in 1980, no CPs were detected in all 51 water samples. In the comprehensive environmental monitoring in 1981, no CPs were detected in all 120 water samples.

The results of the Environmental Investigation on the Status of Pollution by Chemical Substances by the Ministry of the Environment have been reported in *Chemicals in the Environment* published annually in Japan, and LCCPs were selected as the substances analyzed in 2001 [20]. The analysis was conducted according to the analytical method for LCCPs that was reviewed by [21]. CPs analyzed were LCCPs with 40 and 70% chlorine whose standards were available from Wako Pure Chemical Industries, and eight congeners each shown in Table 11 were selected and analyzed. Test samples were collected from water and sediment.

LCCPs were determined by APCI negative ion chemical ionization analysis (APCI-negative) using LC/MS. The detection limits of this method were based on the experimental values obtained in six recovery tests as shown in Table 12.

The environmental survey was conducted at seven sampling sites (three samples/site) and there was no sampling site in which LCCPs were detected in all three samples. LCCPs were detected in only two samples from the Ishikari River estuary and these samples showed different physical appearance in the pretreatment stage.

Table 11 Measured congeners of LCCPs [21]

LCCPs with 40% chlorine	$C_{22}H_{40}Cl_6$, $C_{23}H_{42}Cl_6$, $C_{23}H_{41}Cl_7$, $C_{24}H_{43}Cl_7$, $C_{24}H_{42}Cl_8$, $C_{25}H_{44}Cl_8$, $C_{25}H_{43}Cl_9$, $C_{26}H_{45}Cl_9$
LCCPs with 70% chlorine	$C_{24}H_{31}Cl_{19}$, $C_{25}H_{33}Cl_{19}$, $C_{25}H_{32}Cl_{20}$, $C_{26}H_{34}Cl_{20}$, $C_{26}H_{33}Cl_{21}$, $C_{27}H_{35}Cl_{21}$, $C_{27}H_{34}Cl_{22}$, $C_{28}H_{36}Cl_{22}$

Table 12 Results from the environmental investigation on the status of pollution by chemical substances in Japan (water), monitored in 2001 [20], *n.d.* means not detected

District	LCCPs, chlorine: 40% (Standard detection limit: 0.28 μg L^{-1})	LCCPs, chlorine: 70% (Standard detection limit: 0.14 μg L^{-1})
Ishikari River estuary	0.77, 0.49, n.d.	0.46, 0.83, n.d.
Nagoya port	n.d., n.d., n.d.	n.d., n.d., n.d.
Yokkaichi port	n.d., n.d., n.d.	n.d., n.d., n.d.
Toba port	n.d., n.d., n.d.	n.d., n.d., n.d.
Mizushima offshore (Tamashima)	n.d., n.d., n.d.	n.d., n.d., n.d.
Takamatsu port	n.d., n.d., n.d.	n.d., n.d., n.d.
Kanmon channel	n.d., n.d., n.d.	n.d., n.d., n.d.

Considering a better sensitivity in later years has led to detection of CPs, it is possible to conclude that almost no LCCP was detected in the water.

AIST measured CPs in river water. Samples were collected in Tokyo and Osaka in June 2002. The analytical results are shown in Table 13 and the mean concentrations of SCCPs in the river water samples in the Kanto region was 25.5 ng L^{-1} (Table 14).

Table 13 Analytical results of SCCPs in domestic river water samples (ng L^{-1}) [22]

CPs	Sumidagawa River, Iwabuchi-suimon, Tokyo	Arakawa River, Kasaibashi, Tokyo,	Yodogawa River, Hirakataohashi, Osaka	Yodogawa, Ozeki, Osaka
$C_{10}H_{18}Cl_4$	<5	<5	<5	<5
$C_{10}H_{17}Cl_5$	<5	<5	<5	<5
$C_{10}H_{16}Cl_6$	7.7	8.1	9.5	7.6
C_{10} total	**7.7**	**8.1**	**9.5**	**7.6**
$C_{11}H_{20}Cl_4$	<5	<5	<5	<5
$C_{11}H_{19}Cl_5$	<5 (1.6)	<5 (1.9)	<5 (2.3)	<5 (2.2)
$C_{11}H_{18}Cl_6$	<5 (2.1)	<5 (2.5)	<5 (2.6)	<5 (2.6)
$C_{11}H_{17}Cl_7$	5.6	11	<5 (3.9)	<5 (4.2)
$C_{11}H_{16}Cl_8$	6.5	12	<5 (2.3)	<5 (2.4)
C_{11} total	**12.1 (15.8)**	**23 (27.4)**	**0 (11.1)**	**0 (11.4)**
$C_{12}H_{22}Cl_4$	<5	<5	<5	<5
$C_{12}H_2Cl_5$	<5	<5	<5	<5
$C_{12}H_{20}Cl_6$	<5	<5	<5	<5
$C_{12}H_{19}Cl_7$	<5 (2.2)	<5 (2.1)	<5 (2.0)	<5 (2.1)
$C_{12}H_{18}Cl_8$	<5 (1.0)	<5 (0.9)	<5 (0.7)	<5 (0.7)
C_{12} total	**0 (3.2)**	**0 (3.0)**	**0 (2.7)**	**0 (2.8)**
$C_{13}H_{24}Cl_4$	<5	<5	<5	<5
$C_{13}H_{23}Cl_5$	<5	<5	<5	<5
$C_{13}H_{22}Cl_6$	<5 (0.5)	<5 (0.3)	<5	<5
$C_{13}H_{21}Cl_7$	<5 (0.6)	<5 (0.4)	<5 (0.3)	<5 (0.2)
$C_{13}H_{20}Cl_8$	<5 (0.8)	<5 (0.4)	<5 (0.2)	<5 (0.2)
C_{13} total	**0 (1.9)**	**0 (1.1)**	**0 (0.5)**	**0 (0.4)**
C_{10-13} total	**20 (29)**	**31 (40)**	**9.5 (24)**	**7.6 (22)**

Detection limit is 5 ng L^{-1}. Results in parentheses are data when distinguishable peaks lower than the detection limit are counted

Table 14 Results from environmental investigation on the status of pollution by chemical substances in Japan (sediment) monitored in 2001 [20], *n.d.* means not detected

District	LCCPs, chlorine: 40% (Standard detection limit: 0.038 mg kg^{-1} dw)	LCCPs, chlorine: 70% (Standard detection limit: 0.011 mg kg^{-1} dw)
Ishikari River estuary	n.d., n.d., n.d.	n.d., n.d., 0.013
Nagoya port	0.074, 0.071, 0.057	n.d., n.d., n.d.
Yokkaichi port	0.096, 0.097, 0.34	0.024, 0.076, 0.3
Toba port	2, 0.28, 0.28	0.39, 0.06, 0.064
Mizushima offshore (Tamashima)	n.d., 0.042, 0.045	0.011, 0.033, 0.061
Takamatsu port	0.5, 0.28, 0.47	0.057, 0.041, 0.076
Kanmon channel	0.15, 0.16, 0.83	0.023, 0.049, 0.073

3.3.2 Sediment

EA conducted the environmental monitoring of chemical substances, including all CPs in sediments similar to those conducted in water at 17 sampling sites (three samples/site) in 1980 and the environmental survey at 40 sampling sites (three samples/site) in 1981. CP concentrations were analyzed using GC/ECD with the detection limit of 0.5 mg kg^{-1}.

In 1980, while monitoring the chemical substances in the environment [18], CPs were detected in all three samples from 6 of 17 sampling sites, i.e., Yokohama port (2.7–7.4 mg kg^{-1} dw), Tsurumi River estuary (1.7–3.7 mg kg^{-1} dw), Lake Suwa (1.6–4.8 mg kg^{-1} dw), Osaka port (1.3–4.1 mg kg^{-1} dw), Kobe port (0.9–4.1 mg kg^{-1} dw), and Kure bay (2.0–6.2 mg kg^{1} dw). The highest concentration of six sampling sites was 7.4 mg kg^{-1} dw in Yokohama port and the highest in all samples was 9.4–10 mg kg^{-1} dw in Himeji offshore (Ichi River estuary).

In the 1981 comprehensive monitoring of chemical substances in the environment [19], CPs were detected in all three samples from 8 of 40 sampling sites, i.e., Niigata-Higashi port, Arakawa River estuary, Tama River estuary, Kawasaki port, Yokohama port, Tsurumi River estuary, Kobe port, and Himeji offshore. The highest concentration of eight sampling sites was 8.5 mg kg^{-1} dw in Yokohama port and this concentration was the highest in all samples and was far higher than the other measured values. In the analyses of CPs in fish (edible part) conducted at the same time, no CPs were detected in all 108 samples (28 sampling sites). It should be noted, however, that the detection limit in fish was 0.5 mg kg^{-1} at that time.

In the environmental investigation on the status of pollution [20], LCCPs with 40 and 70% chlorine in sediments were determined by APCI-negative using LC/MS. The detection limits of this method were based on the experimental values obtained in six recovery tests. The results indicated that LCCPs were accumulated in sediments of all sampling sites except the Ishikari River estuary. In the Ishikari River estuary, however, LCCPs were detected in river water but not in sediments, which was the reverse of the result of others.

In the analyses in 1980 [18], all CPs were analyzed although the analytical sensitivity was low. In contrast, in the analyses in 2003 [20], only LCCPs were analyzed. Therefore, it is not possible to directly compare these results. Although the production of CPs reduced by half from 1990 to 2000, considering the high K_{OC} of SCCPs, the low degradation of highly CPs, and a remarkable improvement in analytical sensitivity, it is estimated that SCCP concentrations in sediments have not been reduced significantly, but they possibly gradually decreased over 20 years after 1980.

In the analysis of total CPs (including SCCPs, MCCPs, and LCCPs) conducted by the production plants of CPs in 1978 in sediments around the outlets of wastewater from the plants, it was reported that the concentrations detected ranged from 2.2 to 9.4 mg kg^{-1} ww [2], which were similar to the concentrations in sediments of the Japanese coast that were determined by EA in the same period. It was also reported that the analytical results of total CPs in sediments around the same sampling sites in 2002 (by the same analytical method as used in 1978) were 1.0–3.3 mg kg^{-1} ww [2].

AIST conducted sampling of sediments in two rivers in Tokyo in Kanto region and one river in Osaka in Kansai region in July 2003. The sampling sites were Arakawa River, Tamagawa River in Tokyo, and Yodogawa River in Osaka. The analytical results are shown in Table 15 [22]. The results indicated that the

Table 15 Analytical results of SCCPs in domestic sediment samples (μg kg^{-1} ww) [22]

CPs	Arakawa River, Horikiribashi	Arakawa River, Kasaibashi	Tamagawa River, Oshibashi	Tamagawa River, Den-en Chofu Zeki	Yodogawa River, Hirakata Ohashi	Yodogawa River, Dempo Ohashi
$C_{10}H_{18}Cl_4$	<1	<1	<1	<1	<1	<1
$C_{10}H_{17}Cl_5$	<1	<1	1.1	1.5	<1	12
$C_{10}H_{16}Cl_6$	1.5	4.3	6.8	14	<1	19
$C_{10}H_{15}Cl_7$	2.8	9.4	9.3	21	<1 (0.46)	22
$C_{10}H_{14}Cl_8$	4.1	15	9.4	22	<1 (0.39)	19
$C_{10}H_{14}Cl_9$	3.6	16	5	17	<1	7.1
$C_{10}H_{14}Cl_{10}$	2.1	9.3	2.1	8.1	<1	2.4
$C_{10}H_{14}Cl_{11}$	<1 (0.40)	2.7	<1 (0.79)	1.8	<1	1.3
C_{10} total	**14.1**	**56.3**	**33.7**	**86**	**0**	**82.7**
$C_{11}H_{20}Cl_4$	<1	<1	<1	<1	<1	<1
$C_{11}H_{19}Cl_5$	<1 (0.38)	<1 (0.74)	1.1	1.4	<1	6.8
$C_{11}H_{18}Cl_6$	2.9	5.8	3.7	6.2	<1	11
$C_{11}H_{17}Cl_7$	39	65	13	23	1.3	45
$C_{11}H_{16}Cl_8$	62	106	16	30	2.1	58
$C_{11}H_{15}Cl_9$	47	92	18	43	1.6	41
$C_{11}H_{14}Cl_{10}$	19	49	20	46	<1 (0.48)	11
$C_{11}H_{13}Cl_{11}$	5	9	11	22	<1	<1 (0.75)
C_{11} total	**175**	**326**	**83**	**171**	**5**	**172**
$C_{12}H_{22}Cl_4$	<1	<1	<1	<1	<1	<1
$C_{12}H_{21}Cl_5$	<1	<1	<1 (0.30)	<1 (0.40)	<1	3.0
$C_{12}H_{20}Cl_6$	<1	<1 (0.70)	1.9	3	<1	8.0
$C_{12}H_{19}Cl_7$	<1 (0.80)	3	5.1	8	<1	12
$C_{12}H_{18}Cl_8$	1.3	5	5.2	7	<1	11
$C_{12}H_{17}Cl_9$	1.6	6.3	2.8	4.7	<1	5.5
$C_{12}H_{16}Cl_{10}$	2.3	9.7	2	4.1	<1	2.3
$C_{12}H_{15}Cl_{11}$	1.9	8.6	1.3	2.7	<1	1.0
C_{12} total	**7.1**	**32.6**	**18.3**	**29.5**	**0**	**42.8**
$C_{13}H_{24}Cl_4$	<1	<1	<1	<1	<1	<1
$C_{13}H_{23}Cl_5$	<1	<1 (0.38)	<1 (0.88)	1.2	<1	3.7
$C_{13}H_{22}Cl_6$	1.2	3.5	7.4	13	<1	20
$C_{13}H_{21}Cl_7$	2.9	11	19	30	<1	40
$C_{13}H_{20}Cl_8$	3.4	15	18	27	<1	33
$C_{13}H_{19}Cl_9$	2.4	13	9.1	15	<1	17
$C_{13}H_{18}Cl_{10}$	2.7	14	5.1	7.9	<1	8.7
$C_{13}H_{17}Cl_{11}$	2.8	14	2.8	5.1	<1	2.9
C_{13} total	**15.3**	**69.2**	**61.3**	**98.2**	**0**	**126.4**
C_{10-13} total	**211.1**	**484.4**	**196.6**	**384.7**	**4.9**	**424.0**
Water content	30.2%	52.8%	48.9%	35.3%	19.2%	62.0%
TOC	0.5%	2.5%	2.1%	1.7%	0.05%	4.0%

Detection limit is 1.0 μg kg^{-1} ww. Results in parentheses are data when distinguishable peaks lower than the detection limit are counted concentrations of SCCP with C_{11} was very high, and the concentrations of SCCPs with high chlorine were high in sediments.

3.3.3 Sewage Treatment Plant

AIST measured SCCPs in influents and effluents at three sewage treatment plants (A, B and C) in Tokyo in Kanto region. The processing capacity (average operating rate to processing capacity) of sewage treatment plants A, B, and C in 2001 was 2.71×10^5 m^3 per day (80%), 3.78×10^5 m^3 per day (41%), and 2.02×10^5 m^3 per day (43%), respectively [23]. The analytical results are shown in Table 16 [22].

SCCPs with high chlorine were frequently detected in influents of sewage treatment plants and little difference by the carbon chain length was found. SCCPs with short carbon chain length and low chlorine were released in effluents, as the degradation capacity was low for these congeners. In total, 4.4–13.5% of SCCPs were released into effluents. Effluents from sewage treatment plans are usually diluted with river water approximately tenfold. However, the mean SCCP concentration in effluents from sewage treatment plans in Tokyo was 26 ng L^{-1}, which was close to the SCCP concentrations in the water samples of the Arakawa

Table 16 Analytical results of SCCPs in domestic sewage treatment plant samples (ng L^{-1}) [22]

CPs	STP A influent	STP A effluent	STP B influent	STP B effluent	STP C influent	STP C effluent
$C_{10}H_{18}Cl_4$	<5	<5	<5	<5	<5	<5
$C_{10}H_{17}Cl_5$	6.5	<5	6.5	2.5	11	<5
$C_{10}H_{16}Cl_6$	23	11	34	18	29	9.9
C_{10} total	**29.5**	**11**	**40.5**	**20.5**	**40**	**9.9**
$C_{11}H_{20}Cl_4$	<5	<5	<5	<5	<5	<5
$C_{11}H_{19}Cl_5$	6.2	<5 (2.6)	10	<5 (3.0)	14	<5 (4.1)
$C_{11}H_{18}Cl_6$	15	<5 (3.4)	23	5.3	26	<5 (4.4)
$C_{11}H_{17}Cl_7$	32	7.2	31	9.1	62	5.9
$C_{11}H_{16}Cl_8$	20	7.7	13	<5 (4.3)	47	<5 (3.3)
C_{11} total	**73.2**	**14.9 (20.9)**	**77**	**14.4 (21.7)**	**149**	**5.9 (17.7)**
$C_{12}H_{22}Cl_4$	<5	<5	<5	<5	<5	<5
$C_{12}H_2Cl_5$	<5(2.5)	<5	6.6	<5	8.4	<5
$C_{12}H_{20}Cl_6$	12	<5	26	<5	29	<5
$C_{12}H_{19}Cl_7$	25	<5	31	<5 (2.9)	34	<5 (2.5)
$C_{12}H_{18}Cl_8$	13	<5	13	<5 (1.3)	12	<5 (0.96)
C_{12} total	**50 (52.5)**	**0**	**76.6**	**0 (4.2)**	**83.4**	**0 (3.5)**
$C_{13}H_{24}Cl_4$	<5	<5	<5	<5	<5	<5
$C_{13}H_{23}Cl_5$	<5 (4.5)	<5	5.6	<5	7.1	<5
$C_{13}H_{22}Cl_6$	25	<5 (0.85)	26	<5	33	<5 (0.60)
$C_{13}H_{21}Cl_7$	29	<5 (0.90)	25	<5 (0.19)	31	<5 (0.89)
$C_{13}H_{20}Cl_8$	13	<5 (1.3)	11	<5 (0.79)	12	<5 (0.82)
C_{13} total	**67 (71.5)**	**0 (3.1)**	**67.6**	**0 (0.98)**	**83.1**	**0 (2.3)**
C_{10-13} total	**220 (230)**	**26 (35)**	**260**	**35 (47)**	**360**	**16 (33)**

Detection limit is 5 ng L^{-1}. Results in parentheses are data when distinguishable peaks lower than the detection limit are counted and Sumidagawa Rivers, 31 and 20 ng L^{-1}, respectively. Therefore, the SCCP concentrations in river water in Tokyo are considered extremely high, and it is possible to assume that these concentrations are of worse cases.

3.3.4 Food

Regarding the SCCP concentrations in food, no monitoring data are available in Japan. To supplement the information on SCCP concentrations in food by obtaining the data of food in Japan, a market basket survey was conducted by AIST to measure the SCCP concentrations in food. Based on the results of this survey, the relevance of the estimated concentrations in food is confirmed, and the daily intake of SCCPs in humans is estimated.

Market-basket samples are samples consisting of food items that are purchased in the market representing a typical diet for a certain population. As shown in Table 17, 11 food categories were listed based on 18 food categories used for the National Health and Nutrition Survey conducted by Ministry of Health, Labour and Welfare (MHLW), [22]. Specifically, nuts and seeds with extremely small consumptions were combined with potatoes as one group. Assuming that the SCCP concentrations in sugar and sweets and snacks were similar, these were grouped together, and seasoning and beverages with ND in the analyses by [24] was added to this group. Green and yellow vegetables and other vegetables were grouped together; mushroom and seaweed, of which the consumptions were small, and beans, of which the concentrations of CPs were close to the detection limit in the analyses [24], were added into this group. Seafood, on the other hand was divided into two food categories, as high concentrations of SCCPs were expected in both fish and shellfish.

The intake fractions of food items in each group were estimated from the average intake of whole-generation data of the MHLW's National Health and Nutrition Survey. According to the estimated percentages, foods were purchased in Ibaraki Prefecture and they were mixed to prepare mixed samples. One sample for each food group was prepared. The mixed samples were analyzed using high-resolution gas chromatography and mass spectrometry with electron capture negative ionization (HRGC/ECNI-HRMS). The usage of internal standard of total PCBs was acceptable due to the good recovery percentage; therefore the analytical conditions were found to be ideal. The analytical results of SCCP concentrations in food are shown in Table 18 [22].

3.3.5 Validation of Model Estimation

Analyses of environemntal concentrations of SCCPs were conducted for this assessment in river water and sediment of typical rivers, and the influents and effluents of sewage treatment plants. The scope of monitoring, however, was limited in terms of sampling time and site. As water concentration data usually

Table 17 List of food analyzed by the market basket survey

Food categories	Sub groups		Food analyzed	Origin	Mixed amount (g)
1. Grain crops	Rice	Rice	White rice	Japan	1,445
		Processed rice	Rice threads	Thailand	11.3
	Wheat	Flour	Flour	Japan	28
		Bread	Bread		134
		Pastry	Bean jam bread		34
		Noodles	Raw Udon		34
			Boiled Udon		34
			Raw Soba		34
			Chinese noodles		34
		Dried noodles	Dried Soba	Japan	11
			Macaroni		11
		Instant noodles	Instant noodles		14
	Other crops		Corn starch		8
2. Seeds and potatoes	Seeds		Mixed nuts (peanuts, almonds, walnuts, giant corn, cashew nuts)	China, USA, Peru, India	28
	Potatoes	Sweet potato	Sweet potato	Japan	140
		Potato	Potato	Japan	458
		Other potatoes	Satoimo (taro)	Japan	174
		Potato processed	Konnyaku	Japan	170
3. Sugar, sweets and snacks, seasoning and beverages	Sugar and honey	Sugar	Brown sugar		30
			Honey		7
		Jam	Jam		7
	Snacks	Rice cracker	Rice cracker		9
		Cakes	Baumkuchen		15
		Biscuit	Biscuit		17
		Other snacks	Chocolate		30
			Bean jam rice cake		30
	Seasoning and beverages	Soy sauce	Soy sauce		91
		Sauces	Worcester sauce		23
		Salt	Salt		6
		Other seasoning	Mirin (a sweet sake)		53
		Japanese sake	Japanese sake		66
		Beer	Beer		302
		Other liquors	Liquor	Germany	48
		Other beverages	Cider		263

(continued)

Table 17 (continued)

Food categories	Sub groups		Food analyzed	Origin	Mixed amount (g)
4. Fats	Butter		Butter	Japan	61
	Margarine		Margarine		15
	Salad oil		Salad oil		543
	Animal fat		Animal fat		12
	Mayonnaise		Mayonnaise		293
5. Beans, green vegetables, other vegetables, mushrooms, seaweeds	Beans	Miso	Miso	Japan	36
		Tofu (bean curd)	Tofu (bean curd)	Japan	106
		Tofu processed	Fried bean curd		20
	Carrots		Carrots	Japan	60
	Spinach		Spinach	Japan	48
	Green pepper		Green pepper	Japan	12
	Tomatoes		Tomatoes	Japan	55
	Green vegetables		Leek	Japan	88
			Garland chrysanthemum	Japan	88
			Broccoli	Japan	88
			Parsley	Japan	88
	Daikon (Japanese radish)		Daikon (Japanese radish)	Japan	100
	Onions		Onions	Japan	76
	Cabbages		Cabbages	Japan	65
	Cucumbers		Cucumbers	Japan	31
	Chinese cabbage		Chinese cabbage	Japan	62
	Other vegetables		Long onion	Japan	27
			Myoga (Japanese ginger)	Japan	27
			Ginger	Japan	27
			Gobo (Burdock)	Japan	27
	Pickles		Nozawana (pickled Brassica campestris var. hakabura)	Japan	18
	Mushrooms		Shiitake mushrooms	Japan	39
			Shiitake mushrooms	China	39
	Seaweeds		Wakame seaweed	Japan	13
			Hijiki (Hizikia fusiforme)	Japan	2
6. Fruit	Citrus fruit		Amanatsu (an orange)	Japan	245
	Apples		SunFuji (an apple)	Japan	104
			Ourin (an apple)	Japan	104
	Banana		Banana	Philippines	89

(*continued*)

Table 17 (continued)

Food categories	Sub groups	Food analyzed	Origin	Mixed amount (g)
	Other fruit	Grapes	Japan	284
		Cherry	Japan	83
	Fruit juice	Tomato juice	Turkey, China	91
7. Fish	Salmon and trout	Rainbow trout	Japan	14
		Silver salmon	Japan (culture)	40
	Tuna	Tuna lean meat (bluefish)	Italy	21
		Tuna	Spain	21
		Katsuo (Bonito)	Japan	21
		Mekajiki (swordfish)	Japan	21
	Tai (sea bream) and flat fish	Black flat fish	Japan	50
		Kinnmedai (Alfonsin)	Japan	50
	Aji (horse mackerel or saurel) and Sardines	Maaji (Saurel)	Japan	48
		Gomasaba (Mackerel)	Japan	48
		Samma (Saury)	Japan	48
	Other fish	Koi (Carp)	Japan	60
		Ayu (sweetfish)	Japan	60
	Squid, octopus, crabs	Octopus	Japan	85
		Fire fly squid	Japan	85
	Salted fish	Salted salmon	USA	55
		Salted mackerel	Norway	55
	Dried fish	Shishamo (Smelt)	Canada	31
		Shirasu (young sardine)	Japan	31
		Dried squid	Japan	31
	Canned fish	Tuna		17
	Tsukudani (fish boiled in soy)	Shrimp Tsukudani	Japan	2
	Fish processed	Fried fish (sardine) balls		50
		Chikuwa (fish paste)		50
		Hanpen (fish cake)		50
	Fish sausage	Fish sausage		6
8. Shellfish	Shellfish	Asari (short-necked clam)	Japan	156
		Asari (short-necked clam)	Japan	156
		Scallop	Japan	187
		Hamaguri (Clam)	Japan	26
		Hamaguri (Clam)	China	26
			Japan	145

(continued)

Table 17 (continued)

Food categories	Sub groups	Food analyzed	Origin	Mixed amount (g)
		Shijimi (Corbicula leana)		
		Tsubugai (Buccinidae)	Japan	26
		Akagai (Ark shell)	China	26
	Canned fish	Canned scallop		228
	Tsukudani (fish boiled in soy)	Scallop Tsukudani (fish boiled in soy)	China	24
9. Meat	Pork	Pork		175
		Pork rib		175
	Beef	Beef	USA	131
		Beef	Japan	131
	Chicken	Chicken	Japan	129
		Chicken	Japan	129
	Ham and sausage	Ham		122
	Other meat	Lamb	New Zealand	8
10. Eggs		Quail eggs	Japan	65
		Eggs	Japan	480
		Eggs	Japan	455
11. Milk	Milk	Milk	Japan	832
	cheese	Camembert cheese	Japan	16
	Other dairy products	Cream	Japan	76
		Yogurt	Japan	76

has lognormal distribution, it is assumed that the measured concentrations of SCCPs also have a lognormal distribution. Further, it is assumed that SCCP concentrations in the Tamagawa and Arakawa Rivers, sampling sites in the Kanto region, are distributed according to the distribution of suspended solid (SS) concentrations. The reason for this assumption is that SCCPs easily adsorb on SS. Using a logarithmic mean of the measured concentrations and the logarithmic standard deviation of SS concentrations in the Tamagawa and Arakawa Rivers, a 90% confidence interval is estimated as shown in Table 19.

Regarding the SCCP concentrations in river water and sediment, the estimated values are within the 90% confidence interval, suggesting the relevance of the estimated values. Regarding the SCCP concentrations in the influents and effluents of sewage treatment plants and the release factors to water, the estimated values are close to the means of the measured values, suggesting the relevance of the estimated values.

For risk characterization, the upper limit of 90% confidence interval of the measured concentrations, i.e., 95th percentile are to be used as a worst case. The reasons for this assumption include first the fact that the rivers where samples were collected are located in the Kanto region with many industrial plants. Secondly, the

Table 18 Analytical results of SCCPs in food categories by market basket survey (unit: µg kg^{-1})

SCCPs	Blank	1. Grain crops	2. Seeds and potatoes	3. Sugar, etc.	4. Fats	5. Beans, etc.	6. Fruit	7. Fish	8. Shellfish	9. Meats	10. Eggs	11. Milk
Quantitation limit	0.1	0.1	0.2	0.2	2	0.2	0.2	0.2	0.2	0.5	0.2	0.3
LOD	0.03	0.03	0.06	0.06	0.5	0.06	0.06	0.06	0.06	0.2	0.06	0.1
Lipid content (%)		0.54	0.9	1.86	94.2	0.92	0.04	9.4	0.72	24.7	11.4	6.96
$C_{10}H_{18}Cl_4$	<0.1	<0.1	<0.2	<0.2	<2	<0.2	<0.2	<0.2	<0.2	<0.5	<0.2	<0.3
$C_{10}H_{17}Cl_5$	<0.1	0.14	<0.2 (0.095)	<0.2 (0.12)	<2 (1.1)	<0.2 (0.084)	<0.2	0.28	0.46	0.22	<2 (0.17)	<0.3
$C_{10}H_{16}Cl_6$	<0.1	0.57	0.47	0.42	8.6	0.5	0.39	0.87	2	1.5	0.49	<0.3 (0.25)
$C_{10}H_{15}Cl_7$	<0.1	0.64	0.61	0.75	18	0.47	0.53	2.4	3.9	2	0.45	0.38
$C_{10}H_{14}Cl_8$	<0.1	0.44	0.34	0.5	18	0.31	0.28	4.3	3.6	1.3	0.36	0.37
$C_{10}H_{14}Cl_9$	<0.1	0.13	<0.2 (0.075)	<0.2 (0.17)	9.3	<0.2 (0.082)	<0.2	2.6	1.1	<0.5 (0.35)	<0.2 (0.082)	<0.3 (0.10)
$C_{10}H_{14}Cl_{10}$	<0.1	<0.1 (0.033)	<0.2	<0.2	4.9	<0.2	<0.2	0.88	<0.2 (0.14)	<0.5	<0.2	<0.3
$C_{10}H_{14}Cl_{11}$	<0.1	<0.1	<0.2	<0.2	<2 (0395)	<0.2	<0.2	<0.2	<0.2 (0.10)	<0.5	<0.2	<0.3
C10 total	0	1.9 (2.0)	1.4 (1.6)	1.7 (2.0)	59 (61)	1.3 (1.4)	1.2	11	11 (11)	5.1 (5.4)	1.3 (1.5)	0.75 (1.1)
$C_{11}H_{20}Cl_4$	<0.1	<0.1	<0.2	<0.2	<2	<0.2	<0.2	<0.2	<0.2	<0.5	<0.2	<0.3
$C_{11}H_{19}Cl_5$	<0.1	0.17	<0.2 (0.18)	0.26	<2 (1.8)	0.24	0.28	0.34	0.28	0.65	0.28	<0.3 (0.20)
$C_{11}H_{18}Cl_6$	<0.1	0.16	<0.2 (0.11)	0.22	3.5	<2 (0.18)	<0.2 (0.16)	0.32	0.63	0.53	0.2	<0.3 (0.14)
$C_{11}H_{17}Cl_7$	<0.1	0.24	<0.2 (0.12)	0.21	14	0.21	<0.2 (0.19)	0.94	1.7	0.75	0.2	<0.3 (0.15)
$C_{11}H_{16}Cl_8$	<0.1	<0.1 (0.077)	<0.2	<0.2 (0.094)	8.3	<0.2 (0.10)	<0.2 (0.069)	1.1	1.1	<0.5 (0.30)	<0.2 (0.082)	<0.3
$C_{11}H_{15}Cl_9$	<0.1	<0.1 (0.035)	<0.2	<0.2	4.6	<0.2	<0.2	0.83	0.65	<0.5 (0.30)	<0.2 (0.063)	<0.3
$C_{11}H_{14}Cl_{10}$	<0.1	<0.1	<0.2	<0.2	<2 (0.2)	<0.2	<0.2	0.36	<0.2 (0.15)	<0.5	<0.2	<0.3
$C_{11}H_{13}Cl_{11}$	<0.1	<0.1	<0.2	<0.2	<2	<0.2	<0.2	<0.2	<0.2	<0.5	<0.2	<0.3
C11 total	0	0.57 (0.68)	0 (0.41)	0.69 (0.79)	30 (40)	0.45 (0.73)	0.28 (0.70)	3.8	4.3 (4.5)	1.9 (2.4)	0.69 (0.84)	0 (0.49)
$C_{12}H_{22}Cl_4$	<0.1	<0.1	<0.2	<0.2	<2	<0.2	<0.2	<0.2	<0.2	<0.5	<0.2	<0.3
$C_{12}H_{21}Cl_5$	<0.1	<0.1	<0.2	<0.2	<2	<0.2	<0.2	<0.2	<0.2	<0.5	<0.2	<0.3

(continued)

Table 18 (continued)

SCCPs	Blank	1. Grain crops	2. Seeds and potatoes	3. Sugar, etc.	4. Fats	5. Beans, etc.	6. Fruit	7. Fish	8. Shellfish	9. Meats	10. Eggs	11. Milk
C12H20Cl6	<0.1	<0.1 (0.054)	<0.2	<0.2	<2 (1.9)	<0.2	<0.2	<0.2 (0.078)	0.23	<0.5	<0.2	<0.3
C12H19Cl7	<0.1	<0.1 (0.074)	<0.2	<0.2 (0.072)	12	<0.2 (0.060)	<0.2	0.3	0.65	<0.5 (0.25)	<0.2 (0.079)	<0.3
C12H18Cl8	<0.1	<0.1 (0.061)	<0.2	<0.2	16	<0.2 (0.086)	<0.2	0.34	0.53	<0.5 (0.32)	<0.2 (0.12)	<0.3
C12H17Cl9	<0.1	<0.1	<0.2	<0.2	8.1	<0.2	<0.2	0.25	0.23	<0.5	<0.2 (0.091)	<0.3
C12H16Cl10	<0.1	<0.1	<0.2	<0.2	5	<0.2	<0.2	<0.2 (0.16)	<0.2 (0.11)	<0.5	<0.2	<0.3
C12H15Cl11	<0.1	<0.1	<0.2	<0.2	3.8	<0.2	<0.2	<0.2 (0.083)	<0.2	<0.5	<0.2	<0.3
C12 total	0	0 (0.19)	0	0 (0.072)	45 (46)	0 (0.15)	0	0.89 (1.2)	1.6 (1.8)	0 (0.57)	0 (0.29)	0
C13H24Cl4	<0.1	<0.1	<0.2	<0.2	<2	<0.2	<0.2	<0.2	<0.2	<0.5	<0.2	<0.3
C13H23Cl5	<0.1	<0.1	<0.2	<0.2	<2	<0.2	<0.2	<0.2	<0.2	<0.5	<0.2	<0.3
C13H22Cl6	<0.1	<0.1 (0.059)	<0.2	<0.2	<2 (0.59)	<0.2	<0.2	<0.2 (0.088)	0.29	<0.5	<0.2	<0.3
C13H21Cl7	<0.1	<0.1 (0.084)	<0.2	<0.2 (0.082)	2.8	<0.2	<0.2	<0.2 (0.16)	0.55	<0.5 (0.22)	<0.2	<0.3
C13H20Cl8	<0.1	<0.1 (0.057)	<0.2	<0.2 (0.068)	4.4	<0.2	<0.2	<0.2 (0.17)	0.36	<0.5	<0.2 (0.064)	<0.3
C13H19Cl9	<0.1	<0.1	<0.2	<0.2	2.9	<0.2	<0.2	<0.2 (0.10)	<0.2 (0.12)	<0.5	<0.2	<0.3
C13H18Cl10	<0.1	<0.1	<0.2	<0.2	<2 (1.5)	<0.2	<0.2	<0.2 (0.79)	<0.2	<0.5	<0.2	<0.3
C13H17Cl11	<0.1	<0.1	<0.2	<0.2	<2 (1.2)	<0.2	<0.2	<0.2	<0.2	<0.5	<0.2	<0.3
C13 total	0	0 (0.20)	0	0 (0.15)	10 (13)	0	0	0 (0.60)	1.2 (1.3)	0 (0.22)	0 (0.064)	0
C10–13 total	0	2.5 (3.1)	1.4 (2.0)	2.4 (3.0)	140 (150)	1.7 (2.3)	1.5 (1.9)	16 (17)	18 (19)	7.0 (19)	2.0 (2.7)	0.75 (1.6)
SCCPs concentration in lipid (μg/kg-lipid)		460 (570)	160 (220)	130 (160)	150 (160)	180 (250)	3800 (4800)	170 (180)	2500 (2600)	2.8 (3.4)	1.8 (2.5)	1.1 (2.3)

Table 19 Comparison of measured and estimated values of SCCPs in the environment. Estimated values in river water and sediment are referred from Table 8

Item	Measured value	Estimated value
Concentration in river water	0.020, 0.031 (mean: 0.0255) μg L^{-1} 90% confidence interval: 0.0051– 0.12 μg L^{-1}	0.0375 μg L^{-1}
Concentration in river sediment	0.197, 0.211, 0.385, 0.484 (mean: 0.319) μg kg^{-1} ww 90% confidence interval: 0.060–1.48 μg kg^{-1} ww	0.286 μg kg^{-1} ww

Table 20 Comparison of measured and estimated concentrations of SCCPs in food

Item	Measured value	Estimated value
Green and yellow vegetables and beans	1.7 μg kg^{-1}	2.06 μg kg^{-1}
Seeds and potatoes	1.4 μg kg^{-1}	738 μg kg^{-1}
Meat	7.0 μg kg^{-1}	13.4 μg kg^{-1}
Milk and dairy product	0.75 μg kg^{-1}	4.23 μg kg^{-1}
Fish	16 μg kg^{-1} ww	221 μg kg^{-1} ww

Estimated values are referred from Table 10

SCCP concentrations in river water were approximately half of those in effluents from sewage treatment plants, which indicates that the measured concentrations of river water were extremely high, considering that usual concentrations are about 10% of those in effluents. Thirdly, based on the assumption that the measured concentrations have distribution with the mean of measured concentrations and the standard deviation of SS, the 95th percentile is calculated extremely high. Therefore, it is considered relevant to assume this as a worst case scenario for risk assessment.

Total SCCP concentrations in food obtained through the market basket survey and the estimated concentrations are compared in Table 20. First, the SCCP concentration in the food group including green and yellow vegetables, beans, and others was 1.7 μg kg^{-1}, which is comparable to the estimated concentration in leaf vegetables of 2.06 μg kg^{-1}. Secondly, the SCCP concentration in nuts and seeds and potatoes was 1.4 μg kg^{-1}, which is more than two orders lower than the estimated concentration in root tissue of plants of 738 μg kg^{-1}. The root tissue of plants applied for in the model shoud be fine roots and different from tubers such as potatoes and thick roots like carrots. It was indicated that this difference in the definitions of the root tissue of plants was reflected in the difference in concentrations [25], because the structural characteristics of the model had some effect on the high concentration in root tissue of plants. Therefore, it is concluded that actual concentrations would not be as high as the estimated concentrations.

The SCCP concentration in fish was 16 μg kg^{-1}, which is more than one order lower than the European Union System for the Evaluation of Substances (EUSES) estimated concentration in fish of 221 μg kg^{-1}. The reason for this difference might be that while the subject fish in the model were freshwater species, fish purchased according to the market basket method were primarily marine species, i.e., salmon, tuna, sea bream ,and flounder, and SCCP concentrations in fish living in marine water would be lower than those in river water. Consequently, it is considered that the actual human exposures of SCCPs from fish consumed daily are not so high as estimated.

In contrast, the SCCP concentration in meat was 7.0 µg kg^{-1}, which is comparable to the estimated concentration in meat of 13.4 µg kg^{-1}. The SCCP concentration in milk and dairy products was 0.75 µg kg^{-1}, which is one order lower than the estimated concentration in milk and dairy products of 4.23 µg kg^{-1}.

It was found, however, that SCCP concentrations in fat and oil, which cannot be estimated by the model, were extremely high (140 µg kg^{-1}). The major food items of fat and oil are vegetable oil, mayonnaise, and rapeseed, and the main raw material of these products, was imported from Australia and Canada, so it is indicated that exposure sources may not be domestic.

In summary, the estimated concentrations are not necessarily consistent with the measured concentrations, and therefore, it is not reliable to estimate human intakes of SCCPs via food based on the results of the model estimation. Also, it is assumed that the high concentrations in fat are probably from exposure sources overseas; however, it is not possible to estimate those concentrations by the model. Therefore, for risk characterization in this assessment, it is considered relevant to estimate human intakes directly using the analytical results by the market basket survey as actual intakes of the Japanese.

3.4   Estimation of Indirect Exposure of Humans from the Environment

Inhalation of SCCPs from the atmosphere by humans is small based on the estimation by mathematical modeling, and there are little data about the inhalation toxicity of SCCPs. Thus, SCCP inhalation from the atmosphere is not considered. Based on estimation by mathematical modeling, oral intake of SCCPs via food is large, and therefore, oral intake is estimated using the analytical results of the market basket survey.

The SCCP intake via food is estimated by multiplying the analytical results obtained in the market basket survey with the food consumptions of the Japanese population and dividing by body weight. Food consumption and body weight data are obtained from "amounts of food intake by food group" and "mean body weight and standard deviation" described in the report of the 2000 National Nutrition Survey [26], and probability density function was applied to each parameter in the assumption of lognormal distribution by Monte Carlo simulation. In the market basket survey, fish and shellfish were divided into two groups and analyzed; however, in order to make it consistent with the food categories of the "amounts of food intake by food group", a SCCP concentration in seafood (16.2 µg kg^{-1}) is calculated considering the consumption ratios of fish and shellfish. The estimation results are shown in Tables 20 and 21.

Based on the estimated results, the intake in females aged one year is the highest, with a mean of 0.391 µg kg^{-1} per day and 95th percentile of 0.680 µg kg^{-1} per day.

Table 21 Total SCCP intake in Japanese male by age (unit: µg kg^{-1} per day)

Age (year)	Geometric mean	5th percentile	50th percentile	95th percentile
1	0.371	0.2094	0.3422	0.6741
5	0.2142	0.1171	0.1951	0.3728
10	0.1867	0.1021	0.1804	0.3413
15	0.1374	0.0757	0.1285	0.2429
20	0.1137	0.0594	0.1072	0.2152
25	0.1105	0.0584	0.1049	0.2076
30–39	0.117	0.0629	0.1103	0.2134
40–49	0.1182	0.0648	0.1128	0.2081
50–59	0.1259	0.0676	0.1157	0.214
60–69	0.1159	0.0507	0.0886	0.1821
70<	0.1075	0.0549	0.0972	0.1938

Table 22 Total SCCP intake in Japanese female by age (unit: µg kg^{-1} per day)

Age (year)	Geometric mean	5th percentile	50th percentile	95th percentile
1	0.3911	0.2169	0.3696	0.6795
5	0.2135	0.1187	0.2046	0.3753
10	0.1747	0.0939	0.1664	0.3272
15	0.1284	0.0706	0.1208	0.2233
20	0.1185	0.0638	0.1109	0.212
25	0.1267	0.0697	0.1191	0.2234
30–39	0.1155	0.0605	0.1088	0.2097
40–49	0.1176	0.0641	0.1091	0.2171
50–59	0.1226	0.0676	0.1157	0.2147
60–69	0.0955	0.0507	0.0886	0.1827
70<	0.1039	0.0549	0.0972	0.1952

The market basket method, however, established the percentages of food consumption based on the average intake of whole-generation data of the National Nutrition Survey. For this reason, the possibility cannot be ruled out that the percentages of items in food group are different in one-year old infants from the composition of composite samples in the market basket survey. In the comparison of the food consumption in one-year old infants with the mean in whole-generation, there is little difference in the intake ratios of food items in fat – the food group with the highest contribution. These indicate that the composition of composite samples has little effect on food consumption and it is relevant to estimate the SCCP intake of a one-year old infant by the above method (Table 22).

3.5 Exposure Data Used for Risk Assessment

95-percentile of the measured concentrations is used as a worst case scenario for ecological risk assessment. Estimated values are used for SCCP concentrations in soil, and in water and sediment around release sources, for which measured data are

Table 23 Exposure data used for risk characterization

Risk	Medium	Exposure concentration or intake	Note
Ecological risk (Kanto region)	Water	0.12 μg L^{-1}	Measured value
	Sediment	1.48 mg kg^{-1} ww	Measured value
	Soil	0.150 mg kg^{-1} ww	Estimated value
Ecological risk (local area)	Water	0.0567 μg L^{-1}	SCCP production, estimated value
		0.126 μg L^{-1}	Production of metal working fluids, estimated value
		0.492 (0.265– 0.947) μg L^{-1}	Use of metal working fluids, estimated value
		0.0375 (0.0375 – 0.0375) μg L^{-1}	Manufacturing of SCCP-containing products, estimated value
	Sediment	0.246 mg kg^{-1} ww	SCCP production, estimated value
		0.629 mg kg^{-1} ww	Production of metal working fluids, estimated value
		2.56 (1.36– 4.96) mg kg^{-1} ww	Use of metal working fluids, estimated value
		0.163 (0.163– 0.163) mg kg^{-1} ww	Manufacturing of SCCP-containing products, estimated value
Human health risk (repeated dose toxicity)	Intake via food	0.68 μg kg^{-1} per day	Measured value multiplied by the 95th percentile of intake of 1-year old female
Human health risk (developmental toxicity)	Intake via food	0.223 μg kg^{-1} per day	Measured value multiplied by the 95th percentile of intake of 25-year old female in child-bearing age

not available. Human daily intake of SCCPs is estimated using results of market basket survey, and 95-percentile of intake of one-year old female is applied to repeated dose toxicity, and the 95-percentile of intake of 25-year old female in childbearing age is applied to developmental toxicity. The exposure data used for risk assessment are shown in Table 23.

4 Dose-Response Assessment

Toxicity of SCCPs to ecosystem and human health is summarized in dose-response assessment. The toxicity and bioaccumulation of SCCPs to organisms in the environment are summarized from the existing data of studies in aquatic organisms including fish, and in birds. The percentages of species affected by SCCPs in biotic community in water and sediment are estimated, and the SCCP concentrations as the criteria for screening SCCP effects on ecosystems are established. On the other hand, after reviewing the controversial issues in human health risk assessment of SCCPs in the existing assessments, human health effects of SCCPs are summarized from the existing data of studies in experimental animals including rats and mice, and in vitro studies. The endpoint and no observed adverse effect level (NOAEL) of SCCPs to human health are identified for risk assessment.

4.1 Ecological Toxicity

Through the review of the existing data, it is clarified that for biokinetics of SCCPs in organisms, high CPs are hard to be metabolized and are excreted slowly from the lipid-rich organs. SCCPs are highly bioaccumulative and the main route is identified as uptake of fish from river water through gills, and the BCF is established as 5,900 based on the results of an existing study in Japan.

The ecological toxicity of SCCPs is reviewed in the existing publications of studies on aquatic organisms including fish and in birds, based on highly reliable data as shown in Table 24 where the percentages of species affected by SCCPs in biotic community are estimated. As a result, 5% Hazard Concentration (HC_5), a concentration with which 5% of aquatic species are affected, is estimated from the species sensitivity distribution (SSD) shown in Fig. 7 to be 2.9 μg L^{-1}. This is used as the criteria for the screening assessment for aquatic organisms.

Using equilibrium partitioning, the HC_5 values for sediment- and soil-dwelling organisms are determined to be 11 and 10 mg kg^{-1} ww, respectively. For risk assessment for birds as higher predators, the NOAEL is established as 166 mg kg^{-1}-feed with the endpoint of embryo viability.

4.2 Human Toxicity

4.2.1 Biokinetics

The oral absorption rate of SCCPs is 100-fold higher than that of percutaneous absorption and the rate of gastrointestinal absorption is high. SCCPs are distributed mainly in the organs with high metabolic activity including the liver, kidney, thyroid gland, and adipose and excreted in respiration, urine, and feces. SCCPs with high chlorine content are poorly absorbed, and once absorbed, poorly excreted. With these available data and information without any study having attemped to identify metabolites, it is difficult to identify the specific metabolic pathway of SCCPs. Considering the above, risks are assessed by comparing average human daily intake with NOAELs in this assessment.

Table 24 The lowest NOECs in aquatic species of toxicity studies used for generation of sensitivity distribution of species

Species	CPs	Test method	Concentration, temperature	Solubilizing agent	Endpoint	Test period	NOEC (μg L^{-1})	Reference
Water flea, *Daphnia magna*	C$_{10-12}$, chlorine 58%	Flow-through	3.2, 5.6, 10, 18, 32, 56 μg L^{-1}, 20°C, salinity: 30.5‰	Acetone 67.1 mg L^{-1}	Reproduction inhibition	21 days	5.6	[43]
Mysid shrimp, *Mysidopsis bahia*	C$_{10-12}$, chlorine 58%	Flow-through	0.6, 1.2, 2.4, 3.8, 7.3 μg L^{-1}, 25°C, salinity: 20‰	Acetone (unstabilized)	Death	28 days	>7.3	[44]
Skeletonema costatum	C$_{10-12}$, chlorine 58%	NA	4.5, 6.7, 12.1, 19.6, 43.1, 69.8 μg L^{-1}, 20°C	Acetone 100 μL L^{-1} (79.2 mg L^{-1})	Growth inhibition	96 h	12.1	[45]
Rainbow trout, *Oncorhynchus mykiss*	C$_{10-12}$, chlorine 58%	Flow-through	10°C	Unknown	Sublethal effect	15–20 days	<40	[46]
Bloodworm, *Chironomus tentans*	C$_{10-12}$, chlorine 58%	–	61–394 μg L^{-1}, 21–23°C	Acetone (unstabilized)	Halt of adult emergence	49 days	61	[47]
Sheapshead minnow, *Cyprinodon variegatus*	C$_{10-12}$, chlorine 58%	Flow-through	36.2, 71.0, 161.8, 279.7, 620.5 μg L^{-1}, 25°C, salinity: 25‰	Acetone	Growth inhibition	32 days	279.7	[48]
Selenastrum capriconutum	C$_{10-12}$, chlorine 58%	NA	0.11, 0.22, 0.39, 0.57, 0.90, 1.2* mg L^{-1}, 24°C	Acetone 100 μL L^{-1} (79.2 mg L^{-1})	Growth inhibition	10 days	390	

Values with * are test concentrations exceeding the maximum water solubility of SCCP (0.975 mg L^{-1})

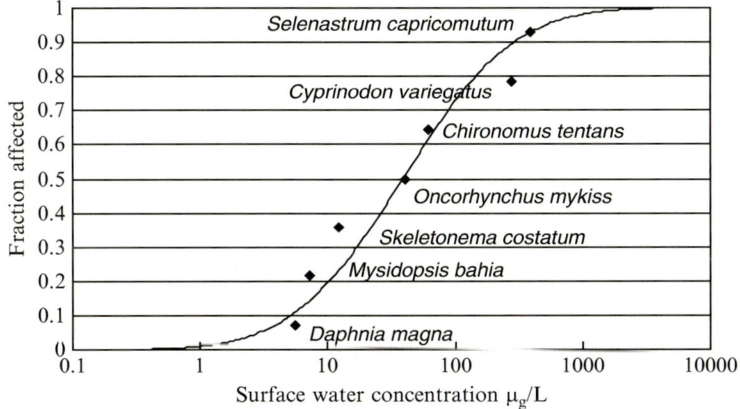

Fig. 7 Species sensitivity distribution (SSD) of aquatic organisms – lognormal distribution

4.2.2 Toxicity to Human Health

Acute toxicity and irritation of SCCPs are low and SCCPs have no potential of sensitization or mutagenicity. Regarding repeated dose toxicity of SCCPs, the results of oral studies in rats and mice indicated that the liver, thyroid gland, and kidney are target organs [27–29].

Marked increases in liver weight and hepatocyte hypertrophy have been shown to be a reflection of peroxisome proliferation [30–33]. Hepatocyte hypertrophy and increased liver weight induced by peroxisome proliferation are known to be specific in rodents [34], and humans are not susceptible to peroxisome proliferation [30, 31]. Therefore, the effects on the liver are considered unlikely to be relevant to human health.

The decrease of T_4 concentration with the increased thyroid weight and enhanced thyroid follicular cell hypertrophy [28, 29] was induced by an increased activity of a liver enzyme (UDPG-transferase) involved in peroxisome proliferation [33, 35, 36]. This mechanism is also specific to rodents. Therefore, thyroidal effects that were observed in the studies in rats and mice are considered unlikely to be relevant to human health.

Regarding the increase in kidney weight, the presence of male rat-specific α2u globulin has not yet been confirmed [37]. This change, however, was observed only in males not in females, and further only in rats. For this reason, it is considered relevant to determine this change to be α2u globulin nephropathy.

Tubular pigmentation was observed in female rats [29]. However, there are no data that rule out the possibility of this effect on humans. Therefore, it is considered reasonable to assume the possible effects on humans. It is assumed that SCCPs accumulated in the tubules have some effects on the kidney due to this toxicity; however, no further information is available at present. Consequently, the NOAEL is established as 100 mg kg^{-1} per day with the endpoint of tubular pigmentation in female rats.

Regarding carcinogenicity, it is relevant that the renal tubular adenoma that was observed in male rats is related to α2u globulin nephropathy, which is specific to male rats, and the possibility that SCCPs are carcinogenic to humans is extremely low.

Regarding reproductive toxicity, the teratogenic effects in fetuses were observed in [38] and the study in rats by [29]; however, it is reasonable to consider that the observed effects are not the direct effects of test substance but the secondary effects derived from the maternal toxicity. In general, a threshold exists for teratogenicity and when some maternal toxicity occurs, fetal anomaly is observed. Therefore, a dose without any maternal toxicity can be used for risk assessment as the no observed effect level (NOEL) for teratogenicity. Consequently, the NOAEL for developmental effect is established as 500 mg kg^{-1} per day.

5 Risk Assessment

In this section, the screening-level assessment of ecological risk to aquatic, sediment- and soil-dwelling organisms are conducted. Human health risk assessment is also conducted. CPs are usually mixtures of different carbon chain length and different degrees of chlorination; therefore, this assessment is concerned with the short-chain length ***(C_{10-13}) CPs as mixtures.

5.1 Ecological Risk Assessment

The methodology of SSD has been used for derivation of environmental quality criteria and for ecological risk assessment. In this chapter, the ecosystem was assumed to be preserved when 95% of species are protected [39], and in this screening-level risk assessment, if the actual or estimated environmental concentration of SCCPs is larger than the HC$_5$ which will protect 95% of species based on the SSD, it is interpreted that it should be assessed further. Risk characterization for birds, the higher predators is also performed because of the high bioconcentration of SCCPs in fish.

5.1.1 Risk Characterization for Aquatic, Sediment-Dwelling, and Soil-Dwelling Organisms

In the screening-level assessment of SCCPs to aquatic, sediment-, and soil-dwelling organisms, risk characterization is conducted to find whether SCCP concentrations are larger than the HC$_5$ estimated from SSD. The data obtained from the SCCP

Table 25 Screening-level ecological risk assessment at regional level

Medium	HC_5	Environmental concentration	Risk characterization
Water	2.9 μg L^{-1}	0.12 μg L^{-1} (the upper limit of 90% confidence interval of measured values)	Concentrations are lower than the HC_5 values, which indicate there are low potential risks
Sediment	11 mg kg^{-1} ww	1.48 mg kg^{-1} ww (the upper limit of 90% confidence interval of measured values)	
Soil	10 mg kg^{-1} ww	0.15 mg kg^{-1} ww (estimated value)	

monitoring in domestic rivers or the estimated values are used as the environmental concentrations of SCCPs. The result is shown in Table 25.

Because the domestic rivers where monitoring was conducted are typical rivers in Japan running through heavily populated areas where many facilities are also located, the measured SCCP concentrations are relatively high. In addition, the SCCP concentrations measured in the effluents from sewage treatment plant were on the same level as the detected concentrations in river water. Therefore, it is relevant to assume an upper limit of 90% confidence interval of the river water data as a worst case scenario in Japan. Screening-level assessment using these upper limits of 90% confidence interval of the monitoring data indicates that the upper limits are lower than the HC_5. As a result, there is a low potential risk to aquatic and sediment-dwelling organisms in the regions.

As there is no monitoring data for soils, the screening-level assessment is conducted using the regional SCCP concentration in soil estimated at 0.150 mg kg^{-1} ww, which is two orders lower than the HC_5. When the distribution of SCCP concentrations in soil is estimated, the possibility is extremely low that the estimated value will exceed the HC_5. This result indicates that there is a low potential risk to soil-dwelling organisms in the regions.

Ecological risks in local areas around plants are also assessed according to the same procedure. There are, however, uncertainties in estimating the SCCP concentrations around release sources in the use of metal working fluids and in the manufacturing process of products containing SCCPs because the number of plants cannot be identified. For this reason, in addition to the estimation of risks based on the estimated release volumes, sensitivity analysis is conducted with the range from half to twice of the estimated release volumes. As a result, in all life stages, environmental concentrations in local areas are lower than the HC_5 and the results of sensitivity analysis do not exceed the HC_5 as shown in Table 26. It is clarified that there is a low potential risk to organisms in local areas around industrial plants.

5.1.2 Risk Characterization for Birds as Higher Predators

Ecological risk to higher predators is a concern because SCCPs are highly bio-accumulative in fish. Risk to birds as higher predators is estimated as the margin

Table 26 Screening ecological risk assessment at local level. All environmental concentrations are estimated values

Medium	HC_5	Life cycle stage	Environmental concentration	Risk characterization
Water	2.9 µg L^{-1}	SCCP production	0.0567 µg L^{-1}	Concentrations are
		Metal working fluid formulation	0.126 µg L^{-1}	lower than HC_5 value, which
		Metal working fluid use	0.492 (0.265– 0.947) µg L^{-1}	indicates there are low potential risks
		Manufacturing of SCCP-containing products	0.0375 (0.0375– 0.0375) µg L^{-1}	
Sediment	11 mg kg^{-1} ww	SCCP production	0.246 mg kg^{-1} ww	
		Metal working fluid formulation	0.629 mg kg^{-1} ww	
		Metal working fluid use	2.56 (1.36–4.96) mg kg^{-1} ww	
		Manufacturing of SCCP-containing products	0.163 (0.163– 0.163) mg kg^{-1} ww	

Values in parenthesis are the range of concentrations when the amounts of SCCP use are estimated as half to twice the median

of exposure ($MOE_{oral,\ fish}$), and MOE is compared with uncertainty factors. $MOE_{oral,\ fish}$ is defined by the following equation:

$$MOE_{oral,\ fish} = \frac{NOAEL}{C_{fish}} \qquad (2)$$

where, the NOAEL has no observed adverse effect concentration in birds of 166 mg kg^{-1}-feed. C_{fish} is a concentration in fish in a local area, i.e., a predator exposure concentration (mg kg^{-1} ww), and $C_{fish,\ j}$ is a concentration in fish for fish-eating predators calculated as the local and regional concentration in river water [16]

$$C_{fish,j} = BCF_{fish} \times (C_{local,j} + C_{reg})/2 \qquad (3)$$

where BCF_{fish} is the BCF in fish as 5,900. $C_{local,\ j}$ is a concentration in river water in a local area j and C_{reg} is a concentration in river water in the Kanto region. EU-TGD [40] proposed to apply an uncertainty factor of 10 for the NOAEL in a reproductive toxicity study, and [41] similarly proposed an uncertainty factor of 10. Therefore, this risk assessment is performed using an uncertainty factor of 10. Using the estimated SCCP concentrations in local environment, risk characterization for birds is performed as shown in Table 27.

As a result, the margins in all life stages are substantially larger than the uncertainty factor of 10, and in the sensitivity analysis also the margins are not

Table 27 Risk characterization for birds as higher predators

Life cycle stage	$C_{\text{local}, j}$	C_{reg}	C_{fish}	MOE
SCCP production	0.0567 µg L^{-1}	0.0375 µg L^{-1}	278 mg kg^{-1} ww	597
Metal working fluid formulation	0.126 µg L^{-1}	0.0375 µg L	482 mg kg^{-1} ww	344
Metal working fluid use	0.492 (0.265–0.947) µg L^{-1}	0.0375 µg L^{-1}	1,562 (892– 2,904) mg kg^{-1} ww	106 (57–186)
Manufacturing of SCCP-containing products	0.0375 (0.0375–0.0375) µg L^{-1}	0.0375 µg L^{-1}	221 (221–221) mg kg^{-1} ww	750 (750–750)

Values in parenthesis are the range of concentrations when the amounts of SCCP use are estimated as half to twice the median

lower than the uncertainty factor of 10. In conclusion, there is no significant risk to birds.

5.2 Risk Characterization for Human Health

Because the major human exposure pathway is food, human exposures to SCCPs as indirect exposures via food are evaluated. The risk of SCCPs to human health is estimated as MOE and MOE is compared with uncertainty factors. The MOE is defined by the following equation:

$$\text{MOE} = \frac{\text{NOAEL}}{\text{DOSE}} \tag{4}$$

where the NOAEL is no observed adverse effect level for human health, and the DOSE is the daily intake for humans. The NOAEL for tubular pigmentation is 100 mg kg^{-1} per day, and a total uncertainty factor of 1,000 is applied; which includes 10 for a study period of less than 1 year; 10 for interspecies differences; and 10 for individual differences. The NOAEL for developmental effects is 500 mg kg^{-1} per day, and a total uncertainty factor of 100 is applied; which includes 10 for interspecies differences and 10 for individual differences.

Based on the food intake estimated with the market basket survey in Japan, the 95-percentile of the intake of a one-year old female of 0.68 µg kg^{-1} per day is applied to the endpoint of tubular pigmentation, and the 95-percentile of the intake of a 25-year old female of child-bearing age of 0.223 µg kg^{-1} per day is applied to the endpoint of developmental toxicity. The result of risk characterization is shown in Table 28.

As a result, although a worst case scenario is considered in this assessment, the MOEs for both toxicity endpoints are larger than the uncertainty factors. In conclusion, there is no significant risk to human health via the environment, and there is no need for further risk assessment based on more detailed exposure assessment.

Table 28 Risk characterization for human health

Endpoint	Tubular pigmentation	Developmental effect
NOAEL	100 mg kg per day (NOAEL in female rats)	500 mg kg per day (NOAEL in female rats)
Uncertainty factor	1,000 (Short-term study × interspecies difference × individual difference)	100 (Interspecies difference × individual difference)
Human intake	0.68 μg kg^{-1} per day (95th percentile of one-year old female)	0.223 μg kg^{-1} per day (95th percentile of 25-year old female in child-bearing age)
MOE	1.5×10^5	2.2×10^6

6 Conclusion

This assessment is developed with the objective of assessing the current situations of ecological and human health risk of SCCPs in Japan based on the exposure and hazard assessments.

Based on the results of exposure and hazard assessments, risk characterization is performed. Screening ecological risk assessment is performed using the HC_5 estimated from SSD. As a result, the 95th percentile of the measured concentrations of SCCPs in water and sediments of general environment is lower than the HC_5, and it is therefore determined that there is low potential ecological risk to aquatic and sediment-dwelling organisms at the regional level. In local areas around plants using metal working fluids, the estimated local concentrations in all stages of life cycle of SCCPs do not exceed the HC_5, and therefore, it is determined that there is a low potential ecological risk. Further, in the risk characterization of birds as the predators of fish with high bioaccumulation of SCCPs, the MOEs in all stages of the life cycle are larger than the uncertainty factor of 10, and therefore, it is determined that there is no significant risk to birds.

Human health risk assessment is performed using the endpoints of tubular pigmentation and the developmental effect. As a result, the MOEs are 1.5×10^5 and 2.2×10^6, respectively, which are larger than the uncertainty factor of 1,000 (short-term study × interspecies difference × individual difference) in tubular pigmentation and that of 100 (interspecies difference × individual difference) in developmental effect. Consequently, it is determined that there is no significant human health risk of SCCPs.

After the result of this risk assessment, all domestic companies in Japan stopped producing SCCPs in 2006, since SCCPs were classified as Class I Chemical Substances Monitored in 2005. In 2009, the import volume of SCCPs is said to be decreasing, but some companies still use SCCPs as metal working fluids, flame retardant, and for other purposes.

References

1. AIST (2008) AIST risk assessment document series No.6, short chain chlorinated paraffins, Advanced Industrial Science and Technology
2. FRCJ (2003) Hearing survey by AIST, flame retardant chemicals association of Japan
3. CMC (1997) Market forecast on flame retardant polymers & additives 1998 (in Japanese)
4. Tsunemi K, Tokai A (2007) Screening risk assessment of short chain chlorinated paraffins in ecosystems using a multimedia model. J Risk Res 10(6):747–757
5. MLIT (2002) White paper on land 2002, Land Information Division/Land and Water Bureau, Ministry of Land, Infrastructure and Transport (in Japanese)
6. IUCLID (2000) IUCLID Dataset, European Chemicals Bureau, European Commission
7. Drouillard KG, Tomy GT, Muir DCG, Friesen KJ (1998) Volatility of chlorinated n-alkanes (C10–C12): vapor pressures and henry's law constants. Environ Toxicol Chem 17(7):1252 1260
8. Drouillard KG, Hiebert T, Tran P, Tomy GT, Muir DCG, Friesen KJ (1998) Estimating the aqueous solubilities of individual chlorinated n-alkanes (C10–C12) from measurements of chlorinated alkane mixtures. Environ Toxicol Chem 17(7):1261–1267
9. Sijm DTHM, Sinnige TL (1995) Experimental octanol/water partition coefficients of chlorinated paraffins. Chemosphere 31:4427–4435
10. Thompson RS, Gillings E, Cumming RI (1998) Short-chain chlorinated paraffin (55% chlorinated): determination of organic carbon partition coefficient. Zeneca Confidential Report BL6426/B
11. Hoechst AG (1988) TA-Luft classification substantialtion by the UCV Department, 20.05.1988 (cited from EU 1999)
12. Hoechst AG (1991) TA-Luft classification substantialtion by the UCV Department, 18.06.1991 (cited from EU 1999)
13. Chemical Substances Council (2004) 34th central environment council, environment and health panel, chemical screening subcommission (in Japanese)
14. Madeley JR, Maddock BG (1983a) The bioconcentration of a chlorinated paraffin in the tissues and organs of rainbow trout (*Salmo gairdneri*), ICI Confidential Report BL/B/2310
15. Madeley JR, Maddock BG (1983b) Toxicity of a chlorinated paraffin over 60 days. (iv) Chlorinated paraffin – 58% chlorination of short chain length *n*-paraffins, ICI Confidential Report BL/B/2203 (cited from EU 1999)
16. EUSES (1996) The European union system for the evaluation of substances, EUSES ver.1.00 user manual
17. SRTI (2004). Japan statistical yearbook 2004, statistics bureau/statistical research and training institute (in Japanese)
18. EA (1980) Chemicals in the environment in 1980, office of health studies, environmental health department, environment agency (in Japanese)
19. EA (1981) Chemicals in the environment in 1981, office of health studies, environmental health department, environment agency (in Japanese)
20. Ministry of the Environment (2003) Chemicals in the environment in 2002, environmental health and safety division, environmental health department, ministry of the environment (in Japanese)
21. Kenmochi T, Nishijima M, Yoshioka T (2000) Survey study of environmental contamination of chemicals, analysis of chlorinated paraffins by LC/MS, Okayama Prefectual Institute for Environmental Science and Public Health annual report 2000: 10–14 (in Japanese)
22. Iino F, Takasuga T, Senthilkumar K, Nakamura N, Nakanishi J (2005) Risk assessment of short-chain chlorinated paraffins in Japan besed on the first market basket study and species sensitivity
23. JSWA (2002) Sewer statistics – government-, FY 2000, Japan Sewage Works Association (in Japanese)
24. Campbell I, McConnell G (1980) Chlorinated paraffins and the environment. 1. Environmental occurrence. Environ Sci Technol 14(10):1209–1214

25. Schwartz S (2000) Quality assurance of exposure models for environmental risk assessment of suvstances, University of Osnabrück, Doctoral thesis
26. MHLW (2002) National nutrition survey 2000, Ministry of Health, Labour and Welfare (in Japanese)
27. ICI (1981) Chlorinated paraffin (58% chlorination of short-chain length n-Paraffins): 14-Day oral toxity study in rats. ICI Central Toxicology Laboratory (Report No. CTL/C/1111)
28. NTP (1986) Toxicology and carcinogenesis studies of chlorianted paraffins (C_{12}, 60% Chlorine) (CAS No. 63449-39-8) in F344/N rats and $B6C3F_1$ mice (gavage studies). Research triangle park, North Carolina, US Department of Health and Human Services, National Toxicology Program (Technical Report Series No.308)
29. Serrone DM, Birtley RDN, Weigand W, Millischer R (1987) Summaries of toxicological data Toxicology of chlorinated paraffins. Food Chem Toxicol 25(7):553–562
30. Ashby J, Brady A, Elcombe CR, Elliott BM, Ishmael J, Odum J, Tugwood JD, Kettle S, Purchase IFH (1994) Mechanistically-based human hazard assessment of peroxisome proliferator-induced hepatocarcinogenesis. Hum Exp Toxicol 13(Suppl 2):S1–117
31. Bentley P, Calder I, Elcombe C, Grasso P, Stringer D, Wiegand HJ (1993) Review section Hepatic peroxisome proliferation in rodents and its significance for humans. Food Chem Toxicol 31(11):857–907
32. Elcombe CR, Bars RG, Watson SC, Foster JR (1995) Draft Paper, accepted for publication by Xenobiotica, Hepatic effects of chlorinated paraffins in mice, rats, and guinea-pigs: species differences and implications for hepatocarcinogenicity
33. Wyatt I, Coutts CT, Elcombe CR (1993) The effect of chlorinated paraffins on hepatic enzymes and thyroid hormones. Toxicology 77:81–90
34. Japanese Society of Toxicology (2003) In: the Educational Committee, the Japanese Society of Toxicology (ed) Toxicology. Asakura Publishing, Japan (in Japanese)
35. ICI draft paper 1: chlorinated paraffins: early time course of thyroid related events in male rats
36. ICI draft paper 2: time course of events in the liver and thyroid of female F344 rats
37. EU (1999) European risk assessment report, alkanes, C_{10-13}, chloro-, European Union
38. ICI (1982) ICI Report CTL/C/1123 (cited from EU 1999)
39. Posthuma L, Suter GW II, Traas TP (2002) Species sensitivity distributions in ecotoxicology, Lewis Publishers, Boca Raton
40. EU (1996) Technical guidance document in support of commission directive 93/67/EEC on risk assessment for new notified substances and commission regulation (EC) No 1488/94 on risk assessment for existing substances
41. OECD (1995) Guidance document for aquatic effects assessment, OECD environment monographs No.92
42. OECD (2004) Emission scenario document on plastics additives, OECD series on emission scenario documents number 3
43. Thompson RS, Madeley JR (1983a) Toxicity of a chlorinated paraffin to *Daphnia magna*, ICI Confidential report BL/B/2358
44. Thompson RS, Madeley JR (1983b) The acute and chronic toxicity of a chlorinated paraffin to the mysid shrimp (*Mysidopsis bahia*), ICI Confidential Report BL/B/2373
45. Thompson RS, Madeley JR (1983c) Toxicity of a chlorinated paraffin to the marine alga *Skeletonema costatum*. ICI Confidential report BL/B/2328
46. Howard PH, Santodonate J, Saxena J (1975) Investigation of selected potential environmental contaminants: chlorinated paraffins, U.S. EPA Report EPA-560/2-75-007
47. EG & G Bionomics (1983) The acute and chronic toxicity of a chlorinated paraffin (58% chlorination of short chain length n-paraffins) to midges (*Chironomus tentans*), Wareham, Massachusettes, EG & G Bionomics, (Report No. BW 83-6-1426). (cited from EU 1999)
48. Hill RW, Maddock BG (1983) Effect of a chlorinated paraffin on embryos and larvae of the sheepshead minnow (*Cyprinodon variegatus*) -study two, ICI Confidential Report BL/B/2327
49. Thompson RS, Madeley JR (1983d) Toxicity of a chlorinated paraffin to the green alga *Selenastrum capricornutum*, ICI Confidential report BL/B/2321

Index

Printed by Printforce, the Netherlands